MODEL
LOCOMOTIVE
CONSTRUCTION

Model
Locomotive
Construction

Martin Evans, M.I.G.T.E.

MODEL & ALLIED PUBLICATIONS LIMITED
13/35 Bridge Street, Hemel Hempstead, Hertfordshire, England

Model and Allied Publications Limited,
Book Division, Station Road,
Kings Langley, Hertfordshire,
England

First Published 1974

© Model & Allied Publications Ltd. 1974

ISBN 0 85242 355 1

PRINTED BY Unwin Brothers Limited
THE GRESHAM PRESS OLD WOKING SURREY ENGLAND
A Member of the Staples Printing Group

Contents

	FOREWORD	vi
	INTRODUCTION	vii
1	TOOLS AND EQUIPMENT	1
2	CHOOSING A PROTOTYPE	7
3	COMMERCIAL MECHANISMS AND FRAME CONSTRUCTION	20
4	MAKING YOUR OWN MOTOR	30
5	CURRENT COLLECTION	45
6	WHEELS AND AXLES, CRANKPINS AND COUPLING RODS	52
7	SPRINGING: BOGIE AND PONY TRUCKS	65
8	CYLINDERS, MOTION WORK AND VALVE GEARS	71
9	MATERIALS AND SUPERSTRUCTURES	82
10	TANK ENGINES	105
11	TENDERS	110
12	DETAIL WORK: RIVETING	118
13	OUTSIDE-FRAME LOCOMOTIVES AND SPECIAL TYPES	141
14	PAINTING AND LINING	151
	INDEX	160

Foreword

It is now well over twenty years since John Ahern's book *Minature Locomotive Construction* was greeted by an enthusiastic public hungry for modelling information and techniques. So efficiently did this little book by the postwar "master" of the art put over its information, that it has remained the locomotive modeller's "bible" ever since, and is still in print today. But 20 years is a long time, and recent years have seen such changes in modelling attitudes and materials that Ahern's publication now seems a trifle dated, with its emphasis on 3rd rail electrification and its lack of information on more modern methods and raw materials, adhesives, etc. What has been needed is a larger, more comprehensive and up-to-date work on precisely the same subject as J. H. A.'s—modelling locomotives in the smaller gauges. Here, at last, is that book.

Few readers will need an introduction to Martin Evans, the author of this book. His long association with live steam locomotives in the large gauges (many model engineers have built locomotives to his designs, which are followed all over the world) has given him an insight into the practical side of locomotive model making which must be quite unequalled. This immensely practical approach shines through every chapter in this book. Martin Evans *understands* locomotives, and brings the thoroughness and meticulous attention to detail of the model engineer to the less demanding world of "0" and "1" gauges. Detail is important. The basic construction of a miniature locomotive body seldom presents many problems, but such subjects as valve gears, and even humble lamp irons and footsteps, require expert guidance, which is given freely here.

JOHN BREWER

Hemel Hempstead,
1973

Introduction

SINCE the end of World War II, there has been some decline in the popularity of the gauge "0" and gauge "1" model railways. This has been caused primarily, I think, by lack of space in the average modern house or flat, the rooms of which, while adequate for a comprehensive "00" or smaller layout, are too small for anything but a very minor "branch line" in the two larger gauges.

This decline was naturally followed by a contraction in the number of models, parts and accessories available from the Trade, and this again reacted further on the interest of the modeller, so that an unfortunate downward "spiral" was set up.

But as so often happens in the modelling world, the decline was halted, and was reversed, with the realisation by many enthusiasts that the larger gauge model offered a greater challenge to the scratch builder, that far more detail could be incorporated in the gauge "0" and "1" model, and that even the space problem could be overcome—by launching out into the garden. The move back to these larger gauges has also been helped by such bodies as the Gauge "0" Guild and Gauge "1" Association.

As far as the building of locomotives is concerned many enthusiasts have been deterred from entering the ranks of the gauge "0" and "1" devotees by the high cost of finished motors or chassis, and by the fact that a lathe of some kind is really essential if such parts as wheels and axles, chimneys, domes and so forth are to be accurately produced.

With the advent of the small and comparatively low-priced model maker's lathe and the fact that many model railway clubs now boast their own workshops, this drawback is greatly diminished, and it is my hope that more and more enthusiasts will attempt that fascinating pastime, the building of gauge "0" and gauge "1" model electrically-driven locomotives.

In writing this book, I have drawn freely on the work of some distinguished model makers, now unhappily no longer with us, notably the late J. N. Maskelyne and the late J. H. Ahern. I also acknowledge much assistance from Mr. H. A. Taylor of Bletchley.

Finally may I record my gratitude to Mr. John Brewer, a former editor of *Model Railway News* for contributing the foreword, and to Mr. D. Pedder for providing the drawings.

MARTIN EVANS

Hemel Hempstead, 1973

Plate 1   A well-known Gauge "0" modeller at his work bench

CHAPTER ONE

Tools and Equipment

ENTHUSIASTS who build small gauge model locomotives are not always so fortunate as to possess a proper workshop or even a permanent bench on which they can work. Many must carry on their model making activities in a living-room, a spare room or even in the kitchen.

While a permanent bench is certainly highly desirable, especially for gauge "1" work, the builder can manage by using a stout board, possibly plywood, and measuring at least 4 ft. × 2 ft., to which a small vice can be screwed. If this portable bench has to be used on a table, the underside should be lined with felt or similar material so as to avoid damage to the surface. Such a portable bench should always have a stout wooden batten screwed to the front edge on the underside, which can engage against the front edge of the table and thus take up the thrust of filing, sawing and similar operations.

Good lighting is, of course, highly desirable for any form of model work, and as much of the work involved in building the bodywork of a model locomotive involves soldering, a convenient socket and plug should be provided for an electric soldering iron.

The author is a strong advocate of a certain amount of silver soldering in connection with the building up of even small gauge models; but this kind of operation, if done in any room other than a proper workshop, is hardly conducive to domestic bliss!

To complete these short remarks on the model locomotive builder's workshop, the reader will appreciate that a more or less permanent form of workshop is very much to be desired. There is one other type of workshop which could be considered and that is what might be called the folding cabinet workshop, which is made on the principle of the bureau with folding top, so that when the bureau is closed the whole structure resembles an item of furniture, thus blending in with the rest of the furniture in the room.

If the enthusiast is able to acquire and accommodate a proper work-bench a convenient size is about 6 ft. long × 2 ft. wide. It should be as rigid as possible and may for preference have a plywood top. A vice should be screwed to the right-hand edge.

Vices

The vice is really one of the most important tools for the model railway man, so some thought should be given to it when purchasing one. It is a mistake to buy too small or too light a vice, even though the work to be done appears very fragile. The author prefers a good quality parallel bench vice with jaws about 3 in. wide and the jaws should have their serrations ground away. Smooth vice jaws of this type are ideal for clamping most materials, though for very soft materials, fibre inserts can be obtained. The vice should be bolted firmly to the top of the bench at such a height that the top of the jaws are just below the operator's elbow and the bench should have a diagonal bracing situated close to the underside of the vice so as to take the thrust of filing and sawing. The vice should also be arranged so that if a long strip or sheet of material is clamped directly between the jaws, it does not foul the edge of the bench and become bent.

Small Tools

Most model railwaymen will possess some good screwdrivers and pliers. The latter should include round-nose, end-cutting and side-cutting. They should be of as good quality as can be afforded. A suitable hammer is a light ball-pein type of about 8 oz. and a small hide or rubber faced hammer is also very useful. Files are most important and their selection should be made with care. The most useful varieties are as follows: an 8 in. flat hand second cut, an 8 in. flat hand smooth, a 6 in. flat hand dead smooth, a 6 in. flat hand smooth, and a 6 in. half-round smooth. Other useful shapes are 4 in. and 6 in. round second cut and smooth files and a few small square and triangular files.

Needle files are extremely useful for model work of all kinds and a selection of these should always be acquired; the most useful shape is probably the half-round, but square, round, triangular and knife-edge should also be obtained.

It is a very good plan to keep a few old files at one's side for use on solder and lead only, while the newer files should be kept for work on brass and nickel-silver.

A small hand-vice with jaws about $\frac{1}{2}$ in. wide and two or three small tool-makers' clamps, say in 2 in. and 3 in. sizes, are also valuable. A hacksaw frame together with blades having thirty-two teeth to the inch is usually found in most amateur workshops and a coping saw is another useful tool for model work, but perhaps the most important tool of this type to the model locomotive man is the jeweller's or piercing saw for which blades can be obtained in a very large variety of sizes. The most useful sizes for the work we have in mind are known as 0000, 000, 00, and 0.

Other essential tools are steel rules, both in 12 in. and 6 in. flexible, marked in both inches and metric, centre punches, pin punches and scribers. The author finds that a slightly blunted scriber makes an ideal centre punch for the finest work where the regulation centre punch tends to hide the exact position which is to be spotted. While a regulation surface plate is not by any means essential, something really flat is desirable when assembling the body-work of locomotives and a piece of thick plate glass about 2 ft. long $\times$ 1 ft. wide is a cheap but quite satisfactory substitute.

Machine Tools

If a locomotive enthusiast can afford both the space and the expense, a small lathe is a tremendous help for a great deal of the work involved in locomotive construction. For locomotives up to gauge "1", a large machine is not necessary and there are several useful machines of moderate price of between $1\frac{1}{2}$ in. and 3 in. centre height which are entirely satisfactory.

Although a small drilling machine is another extremely useful acquisition, where there is not space for both a lathe and a drilling machine, the lathe can always be used for drilling operations whereas the drilling machine makes a very indifferent lathe whatever modifications or additions are made to it. If the locomotive builder is fortunate enough to be able to acquire both a lathe and a drilling machine, a good quality $\frac{1}{4}$ in. capacity precision drill is the one to aim at. If a lathe is acquired, a small power grinder becomes an essential for re-sharpening or grinding the lathe tools. For the smaller lathes using normal tool bits about $\frac{1}{4}$ in. square, a power grinder with wheels about 3 in. diameter is large enough but, of course, if the necessary resources are available, a double-ended 6 in. grinding machine is a very useful addition to the workshop. It should also be remembered that the grinding machine can also be used for polishing or for the attachment of flexible shafts and similar accessories which can be most useful in all kinds of model work.

The small portable electric drilling machines of the type produced by Messrs. Black and Decker, Wolf, Bridges and other firms can also be pressed into service for some model work. In addition to its more normal use, the drill can be secured in the vice or in the special attachment provided by the manufacturers and made to fulfil some of the functions of a lathe for hand turning, the work being mounted in the chuck. It should be realised, how-ever, that used in this way, the portable drilling machine only makes a poor substitute for a proper lathe. Several attachments, however, are supplied by the manufacturers by means of which the drill can be converted into a vertical bench drilling machine, a circular saw and other uses. A drill of this type can also be used to drive a polishing wheel or a small grinding wheel for tool sharpening.

Micrometers and Slide Gauges

A micrometer is always found in the workshop kit of the keen model locomotive man, yet it is not by any means essential as much of its work can be done by a small but good quality slide gauge marked in inches and millimetres with the usual vernier scale. But those who have a lathe will almost certainly wish to acquire a micrometer before very long and here the author feels that is is much more economical in the long run to purchase one of good make such as a Moore and Wright, a Starrett or a Shardlow. Another measuring instrument which has recently appeared on the market is a dial slide gauge reading directly in thousandth of an inch, and this can be obtained in at least three sizes, the smallest of which will be found most useful in small locomotive work.

Screwing Tackle

A few taps and dies are essential for practically any model work and those that should be obtained first are as follows: Second and plug taps in 6, 8 and 10 BA and circular dies $\frac{13}{16}$ in. diameter to suit, plus suitable holders. Sometimes 4 BA taps and dies are also useful especially in gauge "1" work.

The tapping and clearing drill sizes for these taps are as follows: 4 BA— Tapping 33, clearing 27, 6 BA—Tapping 43, Clearing 34; 8 BA—Tapping 50, Clearing 43; 10 BA—Tapping 55, Clearing 50.

Drills and Reamers

Although it is very nice to possess a complete set of number drills, this is by no means essential as it is generally found in practice that only about a quarter of them will be in regular use. Unless expense is not a consideration, the modeller is advised to start off with the more important sizes such as those just given for tapping and clearing holes plus some of the more commonly used inch sizes, i.e. $\frac{3}{32}$ in., $\frac{1}{8}$ in., $\frac{5}{32}$ in., $\frac{3}{16}$ in., $\frac{1}{4}$ in., etc. Other useful sizes are Nos. 74, 72, 70, 68, 66, 64, 62 and 60 which will cover most of the fine work involved in locomotive construction.

Reamers are intended for finishing holes to exact size and if a few are required, a drill a few thou. smaller should be obtained for each size of reamer. But the only reamers which are likely to be wanted in small gauge locomotive work are $\frac{1}{4}$ in. dia., $\frac{3}{16}$ in. dia., $\frac{5}{32}$ in. dia., $\frac{1}{8}$ in. dia. and $\frac{3}{32}$ in. dia.

Soldering Equipment

Although good work can be done with the ordinary copper bit, heated in the fire or over a gas flame, it does not really succeed in the larger sizes and can be quite a nuisance in smaller ones. It tends to over-heat and if small, to cool off too quickly to be of much use. Electric irons are really far more convenient for this work and if funds permit, one of medium size and a small one with a pencil bit should be acquired.

If gas is available, one of the several miniature blowpipes which are now on the market will be found very useful as they give a small but very hot flame. There are also "self-blowing" gas blowpipes which are sufficiently powerful to carry out small silver soldering operations. It is a good plan to devise means to clamp the blowpipe in position so that both hands are left free for the work. In the absence of a gas supply, a small methylated spirit lamp should be used; the ordinary paraffin or petrol blowlamp should be considered as a last resort, though they are generally too fierce for such small work as the building of model locomotives.

Although some builders prefer a paste soldering flux, the author prefers a liquid flux such as Baker's fluid, though this is not to be recommended on models in which tinplate is being used. The important thing to remember in using any soldering flux is to wash the job thoroughly under the tap after soldering work has been completed for the day.

It is very useful to be able to undertake silver soldering when required; the "Easyflo" silver solder manufactured by Johnson Matthey and Co. Ltd. of Hatton Garden, London E.C.1, can be recommended. Easyflo can be obtained in thin strip or wire in many sections and gauge but for small model work, the round wire of 20 or 22 S.W.G. is recommended.

Readers who have not tried their hand at silver soldering may be surprised to know that this is no more difficult than soft soldering. In fact, once a little experience has been acquired, it really requires somewhat less skill on the part of the operator; it does, however, require a much higher temperature. It is not the author's intention to go very deeply into the subject of silver soldering here as this is covered in other recognized textbooks, but suffice it to say that the secret of good silver soldering (and in fact any soldering), is to ensure that the work is absolutely clean.

There is only one way to be really certain of chemical cleanliness and that is to dip the part to be silver soldered for about ten minutes into a solution of dilute sulphuric acid, after which the job is washed, and fluxed with the special Easyflo flux supplied by the makers. It is essential that the job should be heated reasonably quickly and it must also be brought up to a temperature such that the solder itself will melt immediately on touching the work. (This, in the case of Easyflo, means a dull red heat). It is no use at all melting the solder directly in the flame, the job must be hot enough to ensure that the solder will run freely even if the flame is removed for a moment.

Other Workshop Equipment

To conclude these brief notes on equipping the workshop for the building of miniature locomotives, three other items will be found most useful. A magnifying glass, such as the well-known jeweller's glass, one or two pairs of good quality tweezers and some tables giving standard drill and thread sizes and standard wire gauge sizes. The point about the jeweller's eye-glass is that it can normally be held in the eye, thus leaving both hands free for the work, but as some people cannot conveniently hold the glass in their eye a simple form of stand should either be bought or made up.

Finally, a selection of sheets of emery cloth should be acquired and here the author would strongly recommend that the type known as the 3-M-ite or aluminium-oxide cloth should be obtained.

CHAPTER TWO

Choosing a Prototype

THOSE who are not interested in railways are often astonished at the fascination the steam locomotive seems to hold for the enthusiast. It is certainly difficult to explain the extraordinary attraction that the "iron horse" has for its devotees, but perhaps this can be explained by the remarkable liveliness of the steamer and also the astonishing variety of types, classes and designs. The number of different classes of steam locomotives which have been built in Great Britain alone must run into many hundreds and even those who have studied the subject all their lives are continually coming across designs with which they are not familiar.

Another attraction of the steam locomotive, and this applies more particularly to the pre-group era, is the colourful livery which the managements of railways adopted for their engines.

It is difficult to say why some particular class of locomotive appeals much more to the enthusiast than other types; but the author believes that this is connected with the railway and the locomotives with which the railway

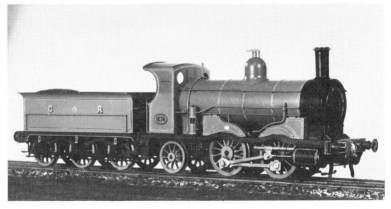

Plate 2   An unusual Gauge "0" locomotive: Caledonian Railway 0–4–2

lover was most familiar during his boyhood. Thus if the enthusiast happened to live beside or near to one of the great main lines, then it is likely that he will always cherish an affection for the locomotives he used to watch go by.

Before deciding on the type of engine the miniature locomotive enthusiast intends to model, he should consider very carefully the pros and cons of the different full size engines and their suitability or otherwise as prototypes.

The complete beginner to this fascinating hobby should obviously choose a fairly small and simple locomotive with inside cylinders. Generally speaking the small tank locomotives, particularly those with a straight footplate and flat cab and bunker sides are much easier to build than those with footplates having various bends or curves in them. Careful note should also be taken of the type of boiler fitted to the prototype, for it will be found that the parallel barrel boiler with a round top firebox is much easier to model than the boiler with a taper barrel or a Belpaire firebox.

Steam locomotives can be classified under five main headings: Express Passenger engines, Mixed Traffic locomotives, Goods or Mineral engines, Tank engines and special types, such as the "Garratt", the "Mallet", the "Fairlie" and so forth.

Express Passenger Engines

4–6–4 type

Only one locomotive of the 4–6–4 wheel arrangement was ever built in Great Britain and even this is sometimes described as a 4–6–2–2. This was the experimental high pressure water-tube locomotive built by the late Sir Nigel Gresley in conjunction with the firm of Yarrow and Co. This particular engine was not successful in traffic and was later re-built into a 3 cylinder locomotive very similar to the streamlined A.4 class after which it performed most satisfactorily.

Owing to the difficult shape of the front of this locomotive, it can only be recommended as a prototype to the highly experienced modeller. Having a long wheelbase with little opportunity to provide much side play for the trailing wheels, a model of this engine would require a rather large radius of curves for satisfactory running.

4–6–2 type

The "Pacific" or 4–6–2 type locomotive was very popular in this country and was also much used in America and other overseas countries. In Great Britain we had the well-known British Railways "Britannia" class and the classes 6 and 8; the different Gresley, Thompson and Peppercorn types of the ex-L.N.E.R., the Stanier "Pacifics" of the old L.M.S. Railway and the

Plate 3   A "Duchess" in Gauge "0" built by Louis Raper

Bulleid streamlined "Pacifics" of the Southern Railway and their re-builds.

The "Pacific" type of engine makes an excellent model for the more advanced enthusiast, although the streamlined L.M.S. and L.N.E.R. "Pacifics" would present great problems in shaping the front of the super-structure. All these prototypes involve the use of rather large radius curved track for satisfactory running and they all involve rather elaborate outside Walschaerts valve gear except for the original Bulleid engines.

4-6-0 type

Locomotives of the 4–6–0 wheel arrangement were extremely popular in Great Britain and as these words are written, there are still a few running on some of the "preserved" lines. The type was used to some extent in the U.S.A. (where it originated) and occasionally on the Continent.

As far as British railways are concerned, the 4–6–0 type has given us many very fine prototypes, both outside and inside cylinder. It has a good wheel arrangement for a medium-sized model, the superstructure allowing the use of a substantial mechanism while the chassis is generally quite flexible with the exception of models of the ex.-G.W.R. four cylinder, "Kings", "Castles" and "Stars". The difficulty with models of these three Great Western types of locomotive is due to the close proximity of the outside cylinders to the rear bogie wheels, a point which will be dealt with in a later chapter.

4-4-2 type

The 4–4–2 or "Atlantic" wheel arrangement makes for a good prototype. Most readers will be familiar with the large "Atlantics" of the old Great Northern Railway and the two classes of "Atlantics" on the late Southern Railway which had round-topped wide fireboxes and trailing wheels with outside frames and axleboxes. The other types of British "Atlantic" locomo-

Plates 4, 5 and 6 Top, a fine Gauge "0" locomotive of the "Star" class. Centre, another view of the same G.W.R. "Star" locomotive. Bottom, a "Royal Scot" class locomotive in Gauge "0", by Louis Raper

*Plates 7 and 8   Top, South Eastern and Chatham Railway, Wainwright 4–4–0,
No. 735, a Gauge "0" model by Mr. W. S. Norris. Below, a fine Gauge "0" Great
Western Railway 2–6–2 tank engine, by L. W. Jones*

tive had narrow fireboxes, most of them having trailing wheels with outside axleboxes, though the old Lancashire and Yorkshire Railway "Atlantics" originally had trailing wheels with inside bearings.

Although the short rigid wheelbase of most of the "Atlantics" could make the type suitable for comparatively sharp curves, this advantage is nullified to some extent by the rigidity of the trailing axle. On the other hand, the superstructure of a model of almost all the "Atlantics" provides ample space for a powerful motor.

4-4-0 type

At one time the 4-4-0 wheel arrangement was the most popular for express passenger work in Great Britain and every major railway company and most of the minor railways possessed several varieties. The classical British 4-4-0 was an inside cylinder machine, though some highly successful engines of this wheel arrangement were built with outside cylinders. Examples are the L.S.W.R. Adams express locomotives and the two very large 4-4-0's built for the Highland Railway, while in more recent years both the L.N.E.R. and the Southern Railway built some most successful three cylinder 4-4-0's— the "Shire" and "Hunt" classes of the late Sir Nigel Gresley and the Southern "Schools".

From the modelling aspect, some of the 4-4-0 locomotives make quite good prototypes; but some of the earlier engines were built with very low pitched boilers, so that those enthusiasts wishing to build them should first check carefully that there is sufficient room inside the superstructure for the mechanism of the model.

The 4-4-0 wheel arrangement for locomotives generally suffers from the disadvantage that it is "front-heavy". This can often be overcome by fairly substantial springing on the leading bogie and by the addition of as much extra weight as possible at the rear end of the locomotive.

2-4-0 type

Towards the end of the nineteenth century, the 2-4-0 wheel arrangement was used to some extent in this country, notably on the old Great Eastern, North Eastern and London and North Western Railways. Most of the 2-4-0's were built for passenger work though some were designed for mixed-traffic service; the majority of them were built with inside cylinders.

From the model enthusiasts' point of view, the 2-4-0 type makes quite a satisfactory engine, most examples having sufficient space in the super-structure for the fitting of the motive power.

Single Wheelers

Most locomotive enthusiasts have a soft spot for the old "single wheelers", once so popular in Great Britain and it is certainly true that some of them were most attractive machines. The famous Dean 4-2-2's of the old Great Western Railway were considered by some the most attractive locomotives ever built, while their performance on light trains was generally excellent. There were also many other most attractive "Singles" in Britain such as the various Johnson types of the old Midland Railway, and the Stirling "eight-footers" of the Great Northern.

Most of the "Singles" make very interesting prototypes and providing that their superstructures are large enough to accommodate the motive power, they are not particularly difficult to build. Some of the "Singles" on the old Brighton Railway had leading and trailing wheels of very large diameter which may make for difficulties in obtaining suitable wheels or wheel castings. It might be thought that a model "Single" would suffer from insufficient adhesion weight, but this is not quite such a drawback as might be expected, as a relatively high proportion of the total weight may be carried by the driving wheels; the type will also generally run quite freely on curves of moderate radius.

Mixed-Traffic Engines

2-6-2 type

There were only a few locomotives built with the 2-6-2 tender type wheel arrangement in Great Britain although it was used quite extensively in America and some other overseas countries. The "Green Arrows" of the old L.N.E.R. were a most successful class of locomotive while this railway also built two lightweight 2-6-2's during the Second World War.

The 2-6-2 wheel arrangement makes quite a large and powerful model, though due to the limitation of the trailing axle, it is not an ideal one for a model railway with sharp curves.

2-6-0 type

The 2-6-0 or "Mogul" type was popular in many countries and several designs were built by the old "Big Four". Nearly all the British "Moguls" had outside cylinders and most were also fitted with Walschaerts valve gear. The L.N.E.R. "Moguls", of which many were built, provide some very large and imposing prototypes which are not at all difficult to make.

It is an excellent type for modelling as the superstructure of practically all the British "Moguls" would be large enough to accommodate a powerful mechanism, and the engine will negotiate fairly sharp curves.

4-6-0 type

The first British 4–6–0 locomotive was designed as a goods engine, the famous Highland "Jones Goods". Later it was developed almost exclusively for express passenger work and then in more recent years 4–6–0's appeared with medium-sized driving wheels for mixed-traffic work. Well-known mixed-traffic 4–6–0's include the Great Western Railway "Halls", "Granges" and "Manors", the L.M.S. "Stanier" 5P5F class, sometimes known as the "Black Fives", the L.N.E.R. Thompson B.1 class and the Great Central B.7 class. There were, of course, many others.

All these locomotives make excellent prototypes for models as they provide a powerful engine which is not too difficult to build and having reasonable curving properties.

Goods or Mineral Engines

0-6-0 type

The 0–6–0 wheel arrangment was at one time the most numerous of all for British locomotives. Almost every railway in this country possessed some goods engines with this wheel layout, the driving wheels ranging from about 4 ft diameter to 5 ft. 8 in. diameter. The 0–6–0 is certainly a most useful type of engine and has aptly been called the "Maid-of-all-Work".

The advantage of the 0–6–0 wheel arrangement is its great compactness and the fact that the whole weight of the engine is available for adhesion, a point which applies equally to the model as well as to the full-sized locomotive. The great variety of the type ensures that every enthusiast can find one to his liking and he may choose, for instance, from the fascinating Webb "Cauliflowers" to the huge Great Eastern J.20 class built by Mr. A. J. Hill at Stratford.

0-8-0 type

Several British railways had 0–8–0 engines and the Fowler locomotives on the L.M.S. and the Webb and Bowen-Cooke designs on the London and North Western were well-known. As with the 0–6–0, the 0–8–0 locomotive allows the whole of the weight of the engine to be used for adhesion. It is not, of course, quite so compact as the 0–6–0 and generally requires a larger radius of curved track for good running. It does, however, make an interesting variation from the typical goods engine and should not be lightly dismissed.

2-8-0 type

The 2–8–0 or "Consolidation" wheel arrangement was first seen in the U.S.A. It achieved great success there and in many other countries. It was

Plates 9, 10 and 11   Top, a simple 0–6–0 saddle tank locomotive by Messrs. Bond's. Centre, a Gauge "0" L.N.W. Railway "Cauliflower", 0–6–0, No. 25 (Mr. Norris). Below, an L.B.S.C. 0–6–0 goods locomotive, No. 429 (Mr. Norris)

first built in Great Britain by the old Great Western Railway where their two very successful classes, the 28XX and 47XX, were well known to all enthusiasts.

The old Somerset and Dorset, Great Northern, Great Central and L.M.S. railways all built successful locomotives with this wheel arrangement. During the 1914–18 War, the Great Central 2–8–0 was chosen by the War Department and was built in large numbers for service in this country and abroad. Again in the 1939–45 War, the L.M.S. Stanier 2–8–0 was preferred by the authorities for war service.

Where the model railway enjoys curves of large radius, the 2–8–0 wheel arrangement is a good one for models as it provides a large and powerful engine which is not too rigid. Greater flexibility can be achieved by removing one or more pairs of flanges from the driving wheels, but this policy should be avoided where possible especially on engines where the wheels are sprung, as will be explained later.

2–8–2 type

This type of locomotive was popular in America and in some overseas countries for many years and in this country two engines were designed by the late Sir Nigel Gresley for the L.N.E.R. for mineral traffic. These two engines proved highly successful and yet strangely enough they had rather a short life. The explanation of this apparent anomaly is that the enormously long trains which they could haul proved too long for the sidings then in existence on the main line from Kings Cross. Gresley also built a few engines of similar wheel arrangement for heavy express passenger work in Scotland. These, too, proved very good engines, but they were not flexible enough for the sharp curves of the Scottish main lines and were accused of spreading the track. They were later re-built by Mr. Thompson, an unfortunate ending for them, in the author's opinion, as they should have been most useful machines further South where the main lines are generally laid to more generous radii.

As far as prototypes for models are concerned, the 2–8–2's clearly call for large layouts with generous curves, while they are really more suitable for construction by the experienced model maker.

Tank Locomotives

Tank locomotives have always been popular in Great Britain especially for semi-fast passenger trains, suburban trains, local goods trains and for shunting. They did not find much favour overseas, however, due to their limited fuel and water capacity.

*Plates 12 and 13   Top, a fine Gauge "1", ex-G.N.R., 0–6–2 tank engine,
built by H. Clarkson and Sons. The model is arranged for two-rail, 24-volt
DC operation. Below, a Gauge "0" South-Eastern and Chatham Railway
0–4–4 tank engine*

One advantage of the Tank locomotive is that the whole locomotive is
self-contained, avoiding all connections between the engine and the tender
and this advantage applies to models also.

The 0–6–0 Tank locomotive, of which there were a great variety on the
railways, provides an excellent prototype for the beginner to model locomo-

tive construction. Some of them, such as the well-known G.6 class of the old L.S.W.R., were most symmetrical and simple in outline and thus easy to model.

Among the larger British Tank locomotives, there were several fine examples such as the various 2–6–2 Tanks of the Great Western Railway, the L.M.S. and the L.N.E.R., the 2–6–4 Tanks of the Southern, L.N.E.R. and the L.M.S. and also the 4–6–4 or "Baltic" Tanks such as the very handsome engines built at Brighton. Another advantage of the Tank locomotive in model form is that even the larger varieties are quite flexible and cause no trouble on curves of medium radius.

Plate 14   A well-made Lancashire and Yorkshire 2–4–2 tank locomotive in "0" Gauge, by Louis Raper

Articulated Locomotives

There were various types of articulated locomotives built by the railways of the world, though only a few were ever seen in Great Britain.

The "Garratt" type was built extensively by the old L.M.S. and one example was also built by the L.N.E.R. It consisted of two complete locomotive chassis, the leading chassis carrying a container for the feed water, the rear supporting the fuel or possibly a combination of fuel and water, the two units being bridged by a large boiler and cab mounted on a girder framework. The "Garratt" type of locomotive proved very popular in South and East Africa and was built for both standard and narrow gauge track. Examples were still running at the time these notes were written.

Other special types of locomotive include the "Fairlie", the "Shay" and the "Heisler". These three types of engine were built mainly for very sharp curves or for lines of a temporary character with severe gradients and possible uneven sections. The "Shay" and "Heisler" locomotives were used to some considerable extent on logging operations especially in North America. Examples of the "Fairlie" can still be seen today on the narrow-gauge Festiniog Railway in North Wales.

Finally, there is the "Mallet" type which was at one time quite popular in the U.S.A. where a very powerful locomotive was required. The main feature of the "Mallet" is the large boiler which supplies two sets of cylinders

Plate 15   A Gauge "0" Rhodesian Railways "Garratt," built by the author. It is fitted with two 12-volt motors, and was exhibited at the Cecil Rhodes Centenary Celebrations, Bulawayo, Rhodesia

with motion and driving wheels, each set of which can swing in relation to the other, thus providing great flexibility. Generally, the "Mallets" were built as compounds, the leading cylinders working on low pressure steam and the rear cylinders on high pressure.

These special types of locomotive are seldom modelled owing to the great amount of work involved, but they should not be overlooked by the more experienced and ambitious railway modeller.

Free-lance Designs

Some model railway enthusiasts prefer to design their own model loco-motives rather than building exact replicas of the full-size engines. In this way, they are able to embody the features of the engine design which most appeals to them. This opens up a completely new field of model locomotive work, but whether the reader decides to enter this field or to stick to the modelling of his favourite prototypes must clearly be a personal decision.

CHAPTER THREE

Commercial Mechanisms and Frame Construction

WE are here concerned solely with electrically driven model locomotives. There are three courses open to the builder of electric models in gauge "0" and "1": he can purchase a "mechanism" from one of the model railway supply houses and build up the remainder of the framework, wheels, etc., following with the superstructure; secondly, he can build a complete chassis for the model, using a ready-made motor and gears, and thirdly he can build the whole of the locomotive chassis himself, buying only those components which are almost impossible for the amateur to make, such as the magnet.

The first method of construction where a commercially-made mechanism is obtained, is not recommended in gauge "0 "and gauge "1". The drawback to this method is that unless the locomotive is a simple 0–4–0, extension frames must be made up and attached either to the superstructure or to the frame plates of the mechanism.

The great majority of locomotive builders are therefore recommended to adopt the second method and to build the complete chassis of the model in one unit, utilising a ready-made electric motor from one of the well-known model railway suppliers.

The first thing to do in building such a chassis is to obtain some hard brass plate of suitable width, $\frac{1}{16}$ in. thick for gauge "0" and either $\frac{1}{16}$ in. or $\frac{3}{32}$ in. thick for gauge "1", and to cut this to the overall length of the whole locomotive, measuring from the back of the front buffer beam to the front of the drag beam. The overall measurement of the motor should be taken very carefully, as should the necessary pitch for the driving gear; that is to say, the centres between the armature shaft and the driving axle. These measurements are then marked lightly on the drawing of the model. Some of the commercial motors are supplied complete with driving axle attached to the motor proper by an enclosed gearbox; in such a case it is only a question of marking out the frames, together with the necessary clearance for the motor.

The centres of the bogie and pony truck wheels and the driving and coupled wheels should be carefully marked out on the side frames of the locomotive and the necessary cut-away to clear the motor and any projections, such as brush gear, which may stand out beyond the magnet or overhang at the sides. The necessary wheel arches for the carrying wheels should next be marked out with a pair of dividers and the top line of the frame where it abuts against the underside of the footplate should be scribed out. All measurements should be taken from the centre of the driving axle, either forwards towards the smokebox end of the model or backwards towards the

Plate 16   A Gauge "0" motor bogie by Messrs. Bond's

cab. Thus the driving axle is treated as a fixed measuring point and every-thing else is considered in relation to it.

The exact cut-away required for the electric motor will depend on its design; for instance, some motors are designed so that the magnet lies between the frames, while in others the magnet is too wide for this and must rest on top of the top edges of the frames.

It is impossible to be absolutely sure that even with the magnet of the electric motor as low down in the chassis as it is possible to have it without fouling the coupled or trailing axleboxes or axles, the top of the magnet is still well clear of the top of the boiler of the locomotive. This will be clear from Fig. 1.

The outline of the locomotive frame plates may next be marked off in accordance with the drawing of the prototype as regards the leading and trailing ends and any holes or cut-aways should be scribed.

Now is the time to decide where to arrange the main cross stretchers of the chassis, not only to support the frames at the correct distance apart, but to act as fixing attachments to the superstructure. It is usual for the chassis to have two holes drilled vertically through the centres of two of the cross stretchers, one close to each end of the frames. The best method of attachment between the chassis and the superstructure is by means of cheesehead screws put from below through clearing holes in these two stretchers into tapped holes in the superstructure. These fixing screws should be at least 6 BA for a "0" gauge locomotive and 5 or 4 BA for a gauge "1"

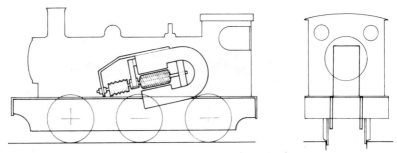

Fig. 1   Fitting the magnet between the frames

model. It will be clear that we cannot tap the footplate itself, as it will be made of quite thin sheet metal; it will therefore be necessary to thicken the footplate locally immediately above the stretcher, in order to provide sufficient thickness of metal to take the screw thread. Alternatively, a standard brass nut may be soldered on top of the footplate. Builders should make quite certain that such a nut or thickening piece goes in a position where it is hidden by the superstructure of the model and does not foul any cross-member.

The advantage of this method of attachment is that should the thread ever become stripped or damaged, a larger drill can be put through from below and the hole retapped a larger size, without having to pull the model to pieces.

One sometimes sees a model locomotive chassis held to the superstructure by one screw only. At the other end there will be a form of housing, probably

Plates 17 and 18   Top, a Gauge "1" chassis for an L.N.E.R.
0–6–2 tank, fitted with a 24-volt DC motor, built by the author.
Below, another view of the same chassis

Plates 19, 20 and 21   Top, the footplate and outside frames of the 7 mm scale Lynton and Barnstaple 2–6–2. Centre, the same model under construction at chassis stage. Below, the boiler and cab added. The model was built by the author

soldered to the back of the buffer beam. A projecting tongue on the end of the chassis engages in this housing. This method is satisfactory for "00" and small gauge models, but is not recommended in gauge "0 "or larger scales.

The holes for the driving and coupled axles should now be drilled, starting with a small centre drill. A decision must now be made as to the type of bearings to be fitted and if these are to be normal press-fit turned bushes, it is simply a question of opening the holes out with care to the size decided upon. For "0" gauge, where the driving axles will probably be $\frac{3}{16}$ in. diameter, the holes should be made $\frac{1}{4}$ in. diameter and if a reamer is available the holes should be reamed after first drilling to a few thou. below the nominal size.

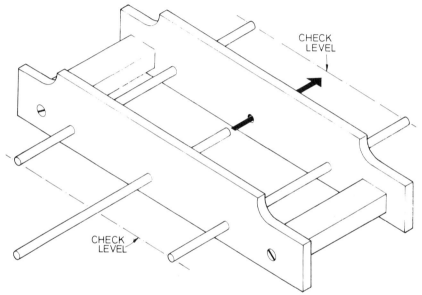

Fig. 2   Checking the axle holes for truth

As it is most important that all the holes for the driving and coupled axles should be exactly in line, it is a good plan, before opening them out, to insert lengths of round silver steel into these holes and check for truth by applying the blade of a square against them. Should the driving axle (in a 6- or 8-coupled model) be very slightly above the level of the others, this will not matter, but should this axle be below the others, which would cause the locomotive to rock on the track, it is essential that this be corrected before any further progress is made. The offending hole can always be shifted slightly by the use of a round file and as the hole will then be made oval, the final sizing drill may not run quite true; provision should therefore be made to ensure that the bush, when fitted, lies at the top of the hole. It could always be soldered to secure it permanently.

We must now determine the position and shape of the insulating block which will be required for the current collectors. The subject of the various methods for current collection for two-rail and three-rail systems will be considered in Chapter Five. Suffice it to say here that a block of bakelite, ebonite or similar insulating material will be required at a suitable position between the driving axle and one of the coupled axles for the conventional methods of current collection. This insulated block may be held in position by two countersunk screws on each side of the chassis, put through clearing holes in the frame, into tapped holes in the block.

At this stage the frame plates may be separated and cleaned up with emery cloth and we are now ready to make the cross stretchers, by which they are to be held together. For most gauge "0" models the overall width over the frame plates is made 1 in. and if the frames are $\frac{1}{16}$ in. thick the stretchers will be $\frac{7}{8}$ in. long. Although the cross stretchers can be made without a lathe, turning is certainly a much better way of making these parts, as one can then ensure that the ends are machined dead square. Although square, or even oblong material is sometimes used for cross stretchers, the author prefers ordinary round brass rod, which makes for a much quicker lathe operation. For gauge "0" locomotive work such round stretchers should be made $\frac{3}{16}$ in. or $\frac{1}{4}$ in. diameter and for gauge "1" $\frac{1}{4}$ in. or $\frac{5}{16}$ in. diameter. Those stretchers which are to carry the fixing screws for the attachment of the chassis to the superstructure, must be made large enough in diameter to take a cross hole for the screws. If the fixing screws are, say, 6 BA (which is approximately 0·110 in. diameter) it will be clear that the stretcher should be at least $\frac{3}{16}$ in. diameter and preferably $\frac{1}{4}$ in. diameter, to ensure that there is plenty of metal around the hole.

The lathe operations for machining a simple round stretcher are as follows: the brass rod is held in the three-jaw chuck (or a collet, if available) and with just a small overhang from the jaws, the end of the rod is then faced off. A centre drill is now put into the tailstock chuck and this is brought up and a centre made in the rod, with the lathe running at a fairly high speed. The centre drill is now changed for a drill of the desired tapping size, which is is then run into the work for the required depth. The tap is put into the tailstock chuck, the tailstock itself is left free to slide along the lathe bed and the tap is now carefully entered by the right hand, while pulling the lathe belt with the left hand.

Using brass rod, it will be found that a "second" tap is quite satisfactory;

it should not be necessary to use taper and plug taps. The tap is next carefully removed by reversing the rotation of the lathe chuck and a countersink is lightly applied to the end of the stretcher to remove burrs. The brass rod is now removed from the chuck and is sawn off at a position slightly beyond the end of the required length. The stretcher is chucked the other way round and faced off once again; it is then removed from the chuck and measured with a micrometer, if available. It will now be seen exactly how much metal is to be removed to bring the stretcher to the exact length required. It is rechucked, centre drilled again and drilled and tapped as before.

If the cross stretcher is one that is to carry a fixing screw, the exact centre of the stretcher should be carefully marked off and deeply popped, when it should be drilled in the drilling machine or lathe. Care should be taken that as the drill meets the cross hole it does not catch up. This can be avoided by holding the stretcher in a small machine vice or any toolmaker's clamp of suitable size.

When all the stretchers required have been completed, the frame plates may be countersunk on the outside to take the fixing screws for the stretchers and a trial assembly carried out. Put all the countersunk screws through their respective holes in the frames and into the stretchers, but for the moment leave all the screws so that they are not quite tightened home and apply the frames to a surface plate or anything really flat, such as a piece of plate-glass when any rock in the plates will immediately become apparent. When the frames are nicely parallel to one another, the stretcher fixing screws are tightened and a check made on the axle holes for parallelism. The quickest way of doing this is to put lengths of round silver steel of the necessary diameter through the axle holes and line them up by eye. Another method is to put the whole chassis on the surface plate once again with these rods of silver steel in position and check the height of each of the rods from the surface plate at the extreme ends, by the use of the surface gauge, calipers or a home-made gauge.

The frame plates may now be dismantled once again and the bushes for the driving and coupled axles turned. For these bushes, a good quality gunmetal or phosphor-bronze should be used. For an axle diameter of $\frac{3}{16}$ in., the bushes may be turned from $\frac{5}{16}$ in. diameter material, being machined down in the lathe to a press-fit for the holes and the frames. For locomotive builders who are the fortunate owners of micrometers, the bushes should be turned down to an interference fit of about $1\frac{1}{2}$ thou. and then eased very slightly with a smooth file to start them in the holes in the frames. Should any of the bushes appear somewhat slack, it is not essential to scrap them, as so long as they are all exactly in line they may be finally secured with a touch of soft solder on the inside. The exact length of the bushes will, of course, depend on the clearance available between the frames and the length of the flange of the bushes which lies on the outside of the frames will depend on the "back-to-back" measurement of the wheels and axles to be used.

Plate 22   Gauge "0" motors fitted with integral gear boxes

Plate 23   Motor units, by Messrs. Bond's, with the gear boxes cut away to show the differing gear arrangements

Plate 24   A Gauge "0" 0–4–2T chassis, built by the author

At this stage, consideration must be given to the exact method of securing the electric motor to the frames. With the type of motor incorporating a gearbox as supplied, for instance, by Messrs. Bond's, the motor will require no fixing at the front end, as it will pivot about the driving axle. At the outer (magnet) end, it may be easily secured by a slotted plate fixed to one of the cross stretchers, the bolt and nut on the end of the magnet holding the motor itself to the slotted plate.

If the motor in use is one without a gearbox, it will depend for its accuracy and good running on a strong attachment at the gear end, as well as at the

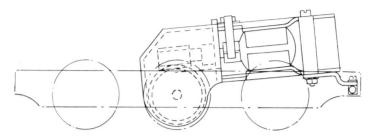

Fig. 3   The electric motor can be secured by means of a slotted plate. The gearbox secures the motor to the driving axle at the front end

magnet end. Sometimes a stout bracket can be bolted or soldered to the overarm of a motor, this bracket having a hole drilled for a bolt, which can be put vertically through this into a tapped hole in a cross stretcher. Vertical adjustment at this end of the motor can then be made very easily by means of washers.

It should now be possible to make a final assembly of the wheels and axles, when the model so far completed may be tried first on the surface plate and then on the track. If any adjustment is required in the location of the axle bushes, this should be carried out by dismantling and giving the necessary attention to the bush which is out of line.

CHAPTER FOUR

Making your own motor

FEW model locomotive enthusiasts attempt the building of their own electric motor for their model locomotive, but for those who are the fortunate owners of a suitable lathe, the work is not as difficult as might be expected.

There is no reason, of course, why certain essential parts of the motor should not be purchased from the Trade, as there are some components in a small electric motor which are rather difficult for the amateur to produce. One of these is the commutator, see Fig. 4. These can generally be bought for quite a small sum from the model supply houses, and other important parts, such as armature stampings, can also be obtained. But the most important component of the motor, that is the magnet (if of the "permanent" variety), will obviously be beyond the scope of the builder.

Plate 25   A Gauge "1" 8-pole, 24-volt motor, by Messrs. Bond's

At one time a suitable magnet could be obtained, made generally in cobalt steel, with the extensions or pole pieces integral with the magnet itself and ground out to form a tunnel embracing the armature. This type of magnet may not now be available, so it is proposed to describe the making up of

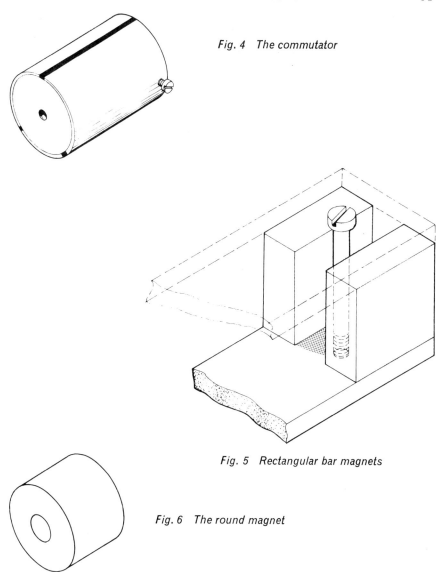

Fig. 4   The commutator

Fig. 5   Rectangular bar magnets

Fig. 6   The round magnet

separate pole pieces, the magnet being a standard rectangular bar type, obtainable ready-made and magnetised from the manufacturers.

If possible, the magnet should be obtained in an alloy such as "Alcomax III", but other magnet alloys are also satisfactory. If the magnet which is to be used is to be of a width equal to the width of the locomotive chassis, it must be provided with a hole through the middle, as all these modern magnet alloys are much too hard to be drilled in their finished state. A round magnet arranged with its axis vertical, may also be considered, provided that it has a hole through the middle for the attachment of the pole pieces, but a third method can be adopted, using two rectangular bar magnets, arranged side by side, at a distance apart to allow for a fixing bolt or screw holding the pole pieces to the top and bottom of the two magnets.

Plate 26   Gauge "0" motor units, by Messrs. Bond's

Pole Pieces

There are several methods of making pole pieces. They can be made from iron castings, the pattern for which may be made from soft wood, such as yellow pine or even deal. Another method is to cut them from solid mild steel, which of course calls for the use of the lathe, while the third method (used in some commercial motors) is to use comparatively thin mild steel strip and shape the outer ends of the pole pieces by pressing to match the diameter of the armature.

If a cast form of construction is decided upon, for the best results the radiused ends of the pole pieces should be machined in the lathe and one way of doing this is by use of a boring tool. The set up is as follows:

The flat end of each pole piece is securely clamped to a small angle plate attached to the faceplate and it is set out from the lathe centre an amount equal to the radius of curvature required. This dimension is important, as upon it depends the clearance between the armature and the pole piece. Although the exact amount of clearance can be adjusted to some extent, when fixing the pole pieces to the magnet or magnets for the purpose of machining the inner surface of the pole piece, it is suggested that a clearance of about 0·015 in. should be allowed. A boring tool is now fitted to the toolpost of the lathe and a very light skim taken across the pole piece to clean up the casting. It will be appreciated that, owing to the sand and scale

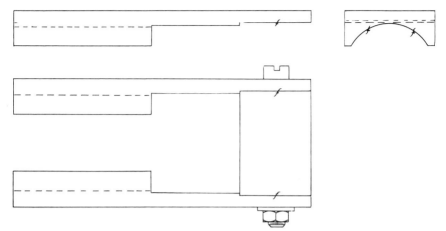

Fig. 7 The casting should be filed around the areas marked f

on the casting, the boring tool will very quickly become blunted, so it may be necessary to remove the tool for resharpening once or twice for each pole piece. Even so, the operation should not take very long, as very little metal has to be removed.

Another method is to use a fly-cutter in the lathe and in this case the pole piece is clamped under the lathe tool holder.

The fly-cutter can very easily be made, using a piece of round silver steel, cross drilled at right-angles near its end and a small round HSS cutter bit put through this hole and clamped by an Allen screw at the end. The cutter bit must be set to describe the diameter required for the armature, plus the necessary clearance between the armature and the pole pieces.

The casting should now be cleaned up by filing on the top, ends and sides and most particularly on that surface which is to be clamped to the magnets. See Fig. 7.

To machine pole pieces from solid mild steel, the following procedure may be adopted:

The rectangular bar should be chosen of the width of the finished pole piece, but of a depth rather more than the finished depth of the pole piece, so as to allow plenty of metal for fly cutting or boring the curved surface. The machining operations will be exactly as described for a cast pole piece. Another method, which is not at all difficult, but involves rather heavy wastage of material, is to choose a length of mild steel bar of such a depth that the whole of the armature tunnel can be machined out of the solid, the pole piece being formed by cutting the bar through the middle on the longitudinal centre line. Having separated the two pole pieces in this way, they are then filed or milled down to the required thickness.

If the pole pieces are to be made as pressings, a soft mild steel strip should be chosen, at least $\frac{3}{32}$ in. thick for a gauge "0" motor and for gauge "1", $\frac{1}{8}$ in. material is recommended. It is not expected that the amateur will be able to make a proper press tool, but a simple type of tool can be made quite easily by choosing two bars of mild steel of suitable width and thickness and shaping each bar to correspond to the top and underside of the desired pole piece. The bar, which has to be hollowed out at a radius to correspond with the armature, can quite easily be machined in the lathe, as described earlier in this chapter, while the matching part of the press tool can be hand filed to the male curvature corresponding to the female. To ensure that the two halves of the tool line up correctly, round dowels should be fitted at each end, clear of the shaped part of the tool.

To operate this improvised press tool a length of material for the pole pieces is inserted between the two halves of the tool: the whole assembly is then put between the vice jaws and the vice tightened.

Using ordinary mild steel as purchased, this simple operation will not, of course, produce the exact radius required for the ends of the pole pieces, owing to the natural spring of the material, but if the steel is heated to a dull red and then slipped into place while still hot, using a pair of pliers, a more correct radius will be produced. Some experiment will obviously be necessary and if it is still found that the material is not being bent to a small enough radius, the two halves of the press tool should be re-machined to a smaller radius.

Fig. 8   Shaping the pole pieces

After successfully shaping the required pair of pole pieces, they should be carefully matched to the shape of the magnet or magnets and drilled for fixing as required. See Fig. 8.

For clamping the two pole pieces to the magnets, it is generally advisable to turn in the lathe a suitably shaped bolt or stud and this should be made to such dimension that an additional length of thread is available at the top for the plate carrying the brush gear and possibly at the other end for attachment of the whole motor to the chassis of the locomotive.

It is sometimes found that it is difficult to keep the pole pieces projecting from the magnet at the right angle (in other words, in line with the armature spindle), but this can be overcome by riveting a short piece of metal, either of square section or angle, across the pole pieces, hard against the edge of the magnet, as shown in the drawing. Where the necessary machining facilities are available, it is not a bad plan to mill away a small section from the thickness of the pole pieces, to ensure an accurate seating for the end of the magnet and at the same time to prevent the pole pieces from shifting.

The Armature Spindle

The next component to consider is the armature spindle and this should be made of silver steel and if the ends are to run in "pin-point" bearings, the coned ends of the spindle should be hardened and polished. The diameter of the spindle should be sufficient to ensure complete rigidity and at least $\frac{3}{16}$ in. diameter is therefore suggested for a gauge "0" motor and $\frac{1}{4}$ in. diameter for a gauge "1".

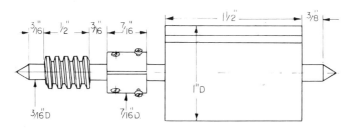

Fig. 9   The armature spindle for Gauge "0"

If ball races are to be used at each end of the armature spindle, the spindle should not be hardened, although there are certain types of pivot ball-bearings or angular-contact races now available, where the end of the spindle bears directly on the races and therefore must be hardened in this case. Suitable miniature ball-bearings or angular-contact ball races can be obtained from such well-known manufacturers as Ina Needle Bearings Limited of Llanelly, South Wales, the Fafnir Bearing Company Limited of Upper Villiers Street, Wolverhampton and the Hoffmann Manufacturing Company Limited of Chelmsford, Essex. All these manufacturers are usually only too pleased to provide necessary technical information regarding the diameters of the spindles and the various tolerances.

Armature Stampings

Armature stampings of the tripolar type can generally be obtained from either the model supply houses or from electrical manufacturers and if only one or two motors are to be made, these should be obtained rather than attempting to make them in the home workshop. As anything up to 50 such stampings may be required for one motor, it would be a very laborious job to cut each one out by hand, so that if the builder cannot obtain the required stampings ready-made, another press tool becomes a necessity.

An important point should be noted about stampings; as received from the suppliers they will almost certainly be very slightly radiused on one side and slightly burred on the other, as they will have been formed in a press tool. Before using them therefore, all the stampings should be filed or rubbed with coarse emery cloth, to ensure that they are quite flat, otherwise when the stampings are assembled the resulting armature will be anything but neat. It will often be found that the hole in the centre of the stampings appears on the large side for the spindle to be used, but providing that the stampings are not too slack when they are all pressed up together on the spindle they will tend to tighten. If, however, they still appear too loose on the spindle, they need not necessarily be discarded, as a good coat of Shellac varnish (or better still Loctite "Hot Strength" adhesive) applied both to the spindle and to the stampings will help to make them a tighter fit on the spindle. When ready for assembly, the stampings are pushed home on the spindle and sufficient stampings should be used to match the length of the pole pieces already made; there is nothing to be gained by making the metallic part of the armature longer than the pole pieces. At each end of the iron stampings a further stamping made from fibre or a similar insulating material should be placed, such stampings being made to the same shape and dimensions as the iron ones. To ensure that the wiring shortly to be placed on the stampings does not "short" to the metallic parts of the armature, suitable electrical insulation should be cut into strips and laid all around the exposed metal parts of the armature, after which the whole of the armature, apart from the spindle itself, should be given a coat of Shellac varnish, or Loctite "Hot Strength" adhesive.

The Commutator

We now require the commutator and this is one component which, as mentioned earlier, is rather difficult to produce in the home workshop. Very likely a suitable commutator may be available from one of the model supply houses or electric motor manufacturers, but if something suitable cannot be obtained, the following method will produce quite a satisfactory commutator of the three pole type. A length of drawn gunmetal or phosphor-bronze of suitable diameter should be obtained and this is chucked in the lathe, centred and bored to a diameter equal to $1\frac{3}{4}$ times the diameter of the spindle. A suitable length of p.t.f.e., nylon or ebonite is now parted off, chucked and turned to a push fit in the hole in the phosphor-bronze. The length required is then dipped in Loctite "Hot Strength" and pushed home. Two thin rings of any suitable metal are now machined to a suitable diameter, so that they can be slipped over the outer ends of the commutator as so far made.

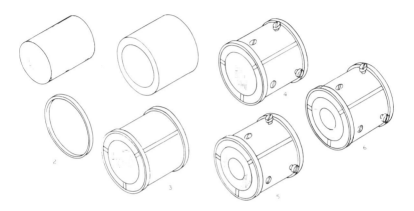

Fig. 10   Making a commutator. 1, A length of suitable metal is bored to $1\frac{3}{4}$ times the diameter of the spindle. 2, Metal rings are made to fit. 3, The nylon or ebonite is fitted and the segments are partly slit. 4, Screws are inserted. 5, Metal rings are removed, segments completely sealed, and the core drilled. 6, Two rings of insulating material are fitted

The next thing to do is to slit the commutator horizontally in three places at 120 deg. around the circumference (that is for a tripolar commutator) using a very thin slitting saw. The slits so formed must not reach quite to either end of the commutator at this stage, otherwise the segments of the metal would almost certainly break away from the insulating core. Small brass or bronze screws are now put in to ensure that the segments of metal cannot come away from the core and for this purpose those which will lie furthest away from the armature should be about 12 BA countersunk and the screws at the armature end of the commutator should be cheesehead and about 10 BA, these screws being to take the ends of the wires from the armature. All these screws are put through clearing holes in the metal segments into tapped holes in the central core and it is, of course, most important that none of these screws are so long that they touch the armature spindle when the commutator has been fitted in place. See Fig. 10. The next stage is to remove the loose metal rings (as these are no longer required) and complete the sealing of the three segments so that they are all well insulated from one another.

The commutator is now chucked in the lathe and set to run as true as possible (using collets if available) and the core is drilled and reamed from the tailstock a light press fit for the armature spindle. To complete the commutator, two fine rings, similar to the temporary ones, but this time made of ebonite, bakelite or a similar insulating material, are turned up and bored a light press fit to go over the outside of the commutator and these are slipped over the extreme ends of the commutator, as shown in Fig. 10.

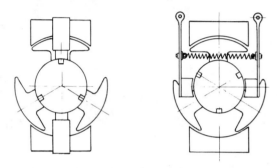

Fig. 11   The arrangement of the commutator slots
in relation to the slots of the armature

The commutator is pressed home on to the armature spindle which is previously given a thin coat of "Loctite" and care must be taken to allow a sufficient gap between the armature stampings and the commutator to allow all the required turns of wire to be wound on the armature. It is, in fact, a good plan to turn up a thin bush of ebonite or fibre and insert this between the armature and the commutator as, in addition to spacing the two correctly, this will act as an additional safeguard against short-circuits between the wiring and the spindle.

The exact radial position of the slots of the commutator in relation to the slots of the armature is most important and this will depend on the location of the brush gear in respect to the pole pieces of the magnet. For instance, if the brush gear is so arranged that the brushes act on the sides of the commutator so that they are at 90 deg. to the vertical axis of the pole pieces, then the slots of the commutator will be in line with the slots of the armature. But if the brushes are arranged vertically, i.e. in line with the axis of the pole pieces, then the slots in the commutator will lie in the middle of the limbs of the armature. See Fig. 11.

Armature Winding

For winding the armature, enamelled copper wire is recommended and the correct gauge of wire can be determined from Table 1, as this depends on

TABLE 1

Recommended Copper Wire Gauges for Armatures
for Model Electric Locomotives

(permanent magnet)

Armature Size	Voltage (D.C.)	Wire Size
½in. dia × ½in. long	12	39 S.W.G.
½in. dia. × ¾in. long	12	38 S.W.G.
1in. dia. × 1in. long	6	28 S.W.G.
1in. dia. × 1in. long	12	32 S.W.G.
1in. dia. × 1in. long	25	34 S.W.G.
1in. dia. × 1¼in. long	12	30 S.W.G.
1in. dia. × 1¼in. long	25	32 S.W.G.
1½in. dia. × 1½in. long	12	26 S.W.G.
1½in. dia. × 1¾in. long	25	28 S.W.G.

the size of the armature, the voltage and the current consumption to be expected. The exact number of turns is best left to the constructor, with the proviso that each limb of the armature should be neatly filled, so that the wires do not project so much that there is a danger of their catching the pole pieces as the armature revolves. It is, however, important that the layers of wire are laid neatly in place on each limb and that there should be the same number of turns on each, otherwise the armature will not run in good balance.

It is a good plan to give the completed armature spindle, with the commutator mounted on it, a rough static test for balance, by mounting it between suitable bearings. It will soon be apparent if one of the limbs of the armature is appreciably heavier than the others. The actual wiring can be done in several different ways, but if only one motor is being constructed the author has found that the operation can be done by hand with the simplest of equipment, if care is taken. The armature should be held by its spindle in the left hand and the right hand used for holding the wire. The wire itself will probably be obtained on a large reel or bobbin and this should be

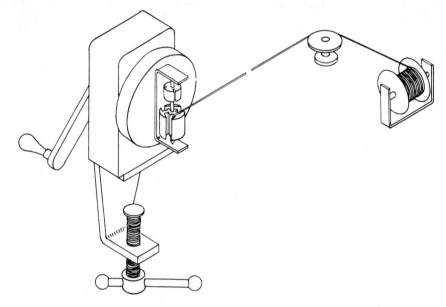

Fig. 12   Winding the armature

mounted on a vertical spindle and arranged so that it is free to revolve, but with a little resistance, as this will help to keep the wire reasonably taut without any danger of its breaking.

Some pressure is necessary between the right hand and the armature itself, to ensure neat wiring and it will generally be found that if the reel is arranged at some distance from the right hand a little pressure can be applied by the hand while the winding is taking place. As for gauge "0" and gauge "1" motors, it is most unlikely that the wire will be thinner than 32 or 34 S.W.G., there will not be much danger of the tension so applied breaking the wire, but this of course is only a matter of experience.

The armature spindle should be held with the commutator away from the operator, the extreme end of the wire should be bared ready for connecting up and a start should be made by winding the first limb in a clockwise direction as steadily as possible. When one limb is completed, a loop is put in the wire, so as to provide sufficient for the second connection, but it must not be broken and wiring of the second limb can then proceed, again in a clockwise direction. After the second limb is completed, another loop is made in the wire and the third limb is wound, again in a clockwise direction; the wire may then be cut off, the free end bared and looped with the start of the wire, so that we now have three loops projecting forwards towards the commutator.

All that now requires to be done is to connect these loops to the cheese-head screws around the commutator. While it is not essential to solder these connections if both the screws and the wire are perfectly clean, a soldered connection is certainly more reliable and for this purpose a low melting point soft solder should be used in conjunction with an electric soldering iron with a fine and clean bit.

Brush Gear

The brushes of any electric motor are very important components and although it is possible to make the actual brushes from copper gauze or similar material, the author recommends that the proper brush material be

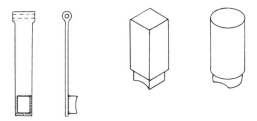

Fig. 13   Two types of brush gear

obtained from the manufacturers. Suitable small copper-carbon brushes can generally be obtained from firms such as the Morgan Crucible Company of Battersea, London, S.W., and these should be obtained copper-plated if it is desired to solder them to the brush arms. Fig. 13 shows two commonly used forms of brush gear, one showing soldered brushes and the other loose brushes working in metal holders. Sometimes cylindrical brushes are adopted, as the holders for these can be made from brass tube, but a rectangular brush is to be preferred for the best performance. Both rectangular and square brass tube in suitable sizes can be obtained from a good metal stockholder, alternatively a suitable rectangular brush holder can be bent up from sheet brass. Another method is to use solid rectangular brass bar, this being drilled and filed out to receive the brush, but this of course involves a good deal more work than the methods first described.

It is important that the brushes of all electric motors are a good fit in their housings, otherwise electrical conductivity will be poor and the springs which bear against the backs of the brushes may become heated and their temper drawn. Another point is that the brush holders, if of the tubular type, should be fairly close to the commutator, so as to give the maximum support to the brush. In this way the brushes will last longer and the electrical efficiency of the motor will be higher.

Tubular brush holders may be fixed in a variety of ways to the main body of the motor. Sometimes the "earthed" brush holder is made a press fit in one of the overarms of the motor or it may be soldered in addition. The brush holder which is to be insulated can be first pressed into a bush made of ebonite, nylon or similar material, this bush in its turn being pressed into the overarm. Where the brush holders are supported from the top overarm only, as in Fig. 14, they may be arranged quite loosely at the top by an

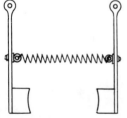

Fig. 14   Brush holders and tension spring

insulated plate and held against the commutator by a bent spring, arranged above or below the overarm, or better still by a small tension spring, arranged between the brush arms just above the commutator; in this case, of course, one end of the spring must be fitted to its respective brush holder by an insulated connection.

Gearing

The type of gearing to be adopted will have been decided upon at an early stage, as upon this depends the type of spindle bearings. The most usual kinds of gears for electric motors with longitudinal spindles are the worm gear and the skew gear. Both these gears set up considerable end thrust, so that plain parallel spindle bearings are unsuitable, either "pin-point" or angular-contact ball races being used.

A suitable gear ratio will depend on the type of locomotive, i.e. whether express passenger, mixed-traffic or slow goods, and for the last named a fairly low ratio should be chosen. While the exact ratio will naturally depend on the revolutions of the armature spindle, the following ratios can be recommended as generally satisfactory:

(a) For express passenger locomotives—Gauge "0" motors: 15–18/1. Gauge "1" motors: 12–15/1.

(b) Mixed-traffic locomotives—Gauge "0" motors: 18–20/1. Gauge "1" motors: 15–18/1.

(c) Freight locomotives—Gauge "0" motors: 25–30/1. Gauge "1" motors: 20–25/1.

Unfortunately, as the freight locomotives have smaller wheels, it is some-times impossible to obtain a low enough ratio with a single pair of gears. This means that a small flat gear or pinion will have to be mounted on an intermediate shaft, such shaft also carrying the worm-wheel or the large skew gear, as the case may be. The pinion then engages with a large diameter flat gear on the driving axle. It should not be forgotten that whatever gear is mounted on the driving axle must clear by a reasonable margin any projec-tions between the running wheels, such as the inside third rail or the studs used in the stud-contact system, which at pointwork protrude above the level of the running rails.

Oil Baths

Some commercial motors have been built with an oil bath enclosing the gearing, so that good lubrication can always be relied upon. The drawback, however, with the oil bath on small motors is the clearance necessary around the final gear, causing a projection which may come dangerously close to the rails. Should an oil bath be required, it is best built up from very thin brass or nickel-silver sheet, the bottom surface of which should be bent around the final gear as closely as possible.

Flywheels

The addition of a flywheel to the armature spindle of a model locomotive electric motor improves the performance of the model on the track quite considerably, giving much smoother starting and stopping. When a motor is fitted with a flywheel the speed controller has to be used with discretion, the power being cut off some distance before the point where the locomotive is required to stop. Unfortunately, many model locomotives which would benefit with the addition of a flywheel appear to have insufficient space in the superstructure to accommodate one; it is, however, quite practicable on most large-boilered engines and in many tank locomotives.

The flywheel is sometimes fitted between the commutator and the outer bearing, but if the design of the magnets allows it, a better position is often at the other end outside the magnets. A small flywheel can sometimes be accommodated in between the armature and the rear bearing, as in Fig. 15.

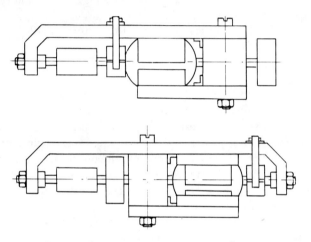

Fig. 15   Alternative positions for the flywheel

It should be remembered when fitting a flywheel to the armature spindle that the addition of the extra weight sets up greater pressures in the bearings, which must therefore be made for the purpose. It should also not be forgotten that as much as possible of the weight of the flywheel should be placed in the periphery rather than in the boss for the best results.

CHAPTER FIVE

Current Collection

THERE are two main types of electrical current collection used in model railways: the three-rail and the two-rail. The former can be sub-divided again into five different methods, as follows—the inside live rail, the outside live rail, the combination of inside and outside live rails, the stud-contact system and finally the overhead wire.

Normally the collectors, whether two-rail or three-rail, must be electrically insulated from the frames and all the metal parts of the model by means of fibre, ebonite, mica, bakelite, nylon, p.t.f.e. or similar material; they are

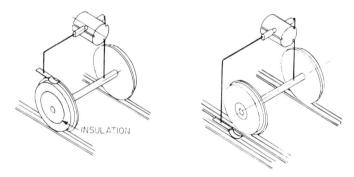

Fig. 16   Two-rail and three-rail collectors

then connected to one of the brushes bearing on the commutator, as in Fig. 16. The wire carrying the current from the collectors to the brush gear should be heavy enough to carry the maximum current which the motor is likely to use, but not so heavy that it is difficult to bend neatly to clear the gears or other obstructions, and it must of course be adequately insulated.

With the normal type of collector for the third rail system, the return path of the current will be via all the driving and coupled wheels of the model and also sometimes through the bogie or trailing wheels. It is some-times suggested that bad electrical contact can be caused by the film of oil which should always be present between the axles and their bearings, but where there are at least four wheels carrying the current in this way, operation is usually quite satisfactory. Should there be any doubt on this point, a simple type of collector, such as hard phosphor-bronze wire or strip, can be arranged to bear directly on the tops of the running rails, but this will not normally be found necessary.

Several types of third rail current collectors are shown in Fig. 17. Type A is a traditional very simple inside collector, generally made of hard phosphor-bronze and it should be made reasonably wide to prevent tilting when the locomotive is running over points. The height of the two "buttons" can be exactly adjusted by the two screws shown, which pass through oval slots in the collector strips into tapped holes in the insulating block.

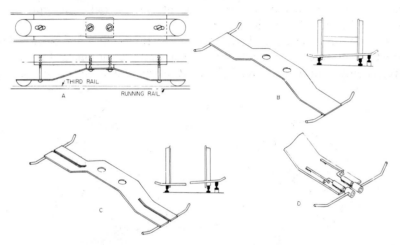

Fig. 17   Third-rail current collectors

Type B is a simple current collector for the outside third rail system, but is not a good one, as it suffers from a tendency to tilt, as can be seen in the drawing. Type C is a better design, for as the collecting bars are separated there is no tendency for the free bar to tilt. The collector seen at D is the best of the three outside collectors, as once properly adjusted it does not easily get out of alignment and both of the bars are quite independent of each other. The bars in this collector can be made of $\frac{1}{16}$ in. diameter nickel-silver or phosphor-bronze for gauge "0" models and $\frac{3}{32}$ in. diameter for gauge "1" models.

The inside current collector shown at A should be as long as possible, subject to clearing the driving and coupled wheels. It will be noted that the tips of the bars of the outside collectors are bent upwards at about 25 deg., so that they come right up on the third rail without catching when approaching it obliquely, as occurs when the locomotive is approaching trailing pointwork, etc.

The apparent decline in the popularity of the third rail system is no doubt mainly due to the unrealistic operation of steam-outline locomotives on electrified track, but the continual adjustments often necessary to the shoes of the collectors may also be a contributing cause. At all events, the two rail system is now much the most popular for the small gauges and is also making great headway in gauge "0" and gauge "1".

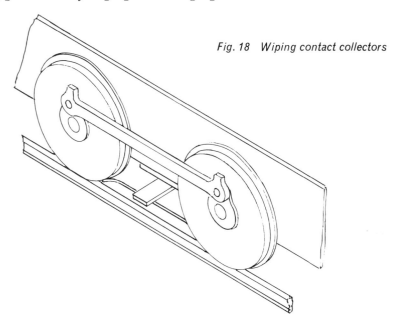

Fig. 18 Wiping contact collectors

Collectors for two rail working are really simpler and easier to make than those for three rail, although against this there is much more work involved in insulating the wheels on one side of the locomotive.

On the insulated side of the model (this is generally but not necessarily the left-hand side) two or more collectors are fitted which make wiping contact with the tread or flange of the wheel, which of course takes the current directly from the surface of the rails. The disadvantage of collecting from the back of the flange, as is sometimes done, is that the wheels are pushed slightly to one side of the frames according to the amount of end play allowed in the axles. Quite a satisfactory two rail collector can be made from round nickel-silver or phosphor-bronze wire, which can be soldered to a projection standing out between the wheels from a small plate secured to the insulating block. While this type of collector must constitute a slight brake on the wheels, it does tend to keep reasonably clean in service.

A direct track collector, where the collecting shoe rubs on the railhead itself, eliminates one possible source of bad electrical contact, but on the other hand it does tend to become dirty and must, in any case, be duplicated for reliable running. This type of collector is so simple that little more need be said about its construction, but its location should be arranged within the fixed wheelbase to ensure correct alignment with the railhead.

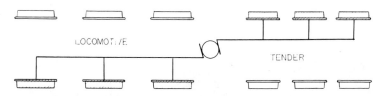

Fig. 19　The circuit through locomotive and tender wheels

One advantage of the two rail system is that collectors can be dispensed with altogether. This can be done on tender locomotives, by the simple device of insulating the locomotive wheels on one side and those of the tender on the other. In this system current passes through the earthed wheels of the locomotive via the motor and from the insulated wheels on the opposite side of the tender. The circuit is illustrated in Fig. 19. in which the wheels shown shaded are the earthed ones and are in electrical contact with the frames of their respective units. A flexible lead is taken from the insulated brush of the commutator to the earthed frame of the tender by means of a fastener or similar device. It will be seen that the locomotive and tender must be insulated from each other, but this is not difficult to arrange. The drag beam of the locomotive can be made of fibre or bakelite and the front beam of the tender can also be made of a similar material, to prevent possible short-circuits on sharp curves.

This collection system can be used on tank locomotives if there is a bogie which can be insulated completely from the frames and can thus be used to provide the return path. With this arrangement it becomes rather difficult to guard against accidental contacts, but in most cases it is possible to fix a sheet of fibre or other insulating material between the frames to prevent possible short-circuits. Another solution would be to make the portion of the frames to which the bogie is attached of an insulating material. It is doubtful whether the system would work successfully on locomotives fitted only with two-wheeled trucks.

The Stud Contact System

The stud contact system of current collection has become quite popular in recent years, especially in garden railways. As in the conventional three rail system, none of the wheels of the locomotives or rolling stock have to be

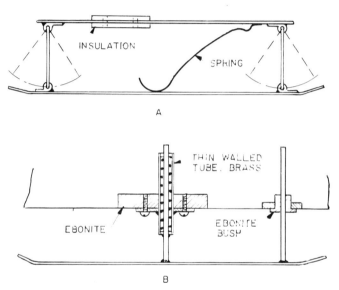

Fig. 20 The stud contact system. A, the pantograph principle. B, Vertical support and guide

insulated, yet the track is not spoilt by the addition of a third rail, which may be completely out of place on a "steam-outline" railway.

The only possible objection to the stud contact system is that it requires a rather elaborate form of current collector and the mounting of this on locomotives, especially those with small wheels, often presents considerable difficulty. The trouble is that the collector must be long enough to bridge the interval between any two studs to ensure continuity of current supplied and it must also be able to rise and fall in a strictly vertical plane to accommodate itself to studs of different heights. Thus it cannot be allowed to tip fore and aft when moving from a high stud to a lower one or vice versa.

There are two main types of collectors for the stud contact system. One is based on the pantograph principle, while the other relies on a vertical support in the centre of the collector, with an additional vertical guide located at one end. This means that there must be two guides projecting upwards into the boiler of the model.

The stud contact collector made on the pantograph principle dispenses with any guide rods. Everything is below the chassis frame as with a normal three rail collector and it can be fitted to practically all model locomotives, though it may have to be modified to suit the particular case. The only slight drawback to the pantograph type is the possibility that the collector or "skate", as it is sometimes called, may foul the final gear wheel on the driving axle when it is at its highest position, as for instance when the locomotive is running over pointwork.

The construction of the pantograph type of collector is as follows: A thin metal plate $\frac{5}{8}$ in. to $\frac{3}{4}$ in. wide (for a "0" gauge locomotive) is attached to the underside of the locomotive frame by two screws, but insulated from it by a strip of fibre or similar sheet material. On the undersurface of this plate two double arms or stirrups are mounted, made of thin hard wire such as nickel-silver or stainless steel and free to swing in a fore and aft direction, carrying the "skate" of the collector, the "skate" being of similar width to the upper plate. Thus the "skate" can swing to a limited extent either towards the

Plate 27   A stud-contact skate on a Gauge "0" 0–4–2T chassis

front or rear of the locomotive when passing over studs of varying height, yet it will remain at all times horizontal and parallel to the surface of the track. It is maintained in tension by a spring of some sort, generally a leaf spring made of hard phosphor-bronze. A lead is then taken from the upper surface of the "skate" to the insulated brush of the motor. It should be noted that this lead should not be attached to the upper plate, owing to possible bad electrical contact through the arms or stirrups.

As mentioned earlier, difficulties may sometimes arise in finding room for the collector where it will be out of the way of the axles and the final gear wheel. In the case of tender locomotives, it is often a good scheme to mount the stud contact collector underneath the tender and take the current to the motor by a flexible lead underneath the cab fall plate. The tender may require some extra weight added to it, to give the best results. In tank locomotives it is often necessary to cut a slot in the upper mounting plate for the final gear wheel to project through.

The mounting plate, which is attached to the locomotive, can of course be made shorter than the "skate" itself, but this should not be done unless it is essential to clear some obstruction on the model, for if the pantograph base is relatively short and the ends of the "skate" project some distance beyond it, there is likely to be a loss of accuracy. The pantograph base should always be as long as possible, as it should be remembered that the length of the "skate" must be sufficent to ride over enough studs in the track to give easy running. It is a common mistake in stud contact work to place the studs too far apart and this is often not discovered until some goods locomotives with small driving and coupled wheels have been built or obtained for running on the railway.

CHAPTER SIX

Wheels and Axles, Crankpins and Coupling Rods

Wheels

Although it is possible to obtain die-cast locomotive wheels ready-made from the Trade for gauge "0" models, the author believes that they are not available for any of the larger gauges. But this is not in any way a disadvantage, because from gauge "1" upwards either cast iron wheels or cast wheel centres with steel rims are much to be preferred.

Plate 28   A 7 mm scale model L.B.S.C.R. 2–2–2 engine, built by the late J. N. Maskelyne in 1933, constructed to test the feasibility of true-scale wheels

One well-known Trade firm supplies both iron castings for a complete range of gauge "0" locomotive wheels and also the finished wheels with ready-made axles to suit, the driving and coupled wheels being supplied ready drilled and tapped for the crankpins. The method used by this Company for attaching the driving and coupled wheels to the axles at the right angle in relation to one another ("quartering") is to machine the outer ends of the axles to a square section for a width of about $\frac{1}{16}$ in. and then to thread the remainder of the axle to take a circular brass nut, which after assembly fits into a recess in the outside of the wheel boss. The back of the driving or coupled wheel is recessed in a square shape to match the squared end of the axle, thus it is very easy to arrange for the driving wheels on one side of the model to be at right angles to those on the other side.

If the builder intends to turn his own cast iron locomotive wheels, he will probably find this squared fixing too difficult and it is therefore suggested that he adopt a similar method to that used in full-size practice, i.e. to machine the wheels to a press fit on the turned ends of the axles and to arrange for the "quartering" to be done before final assembly.

The press fit method is perfectly satisfactory in any gauge from "0" upwards and in fact it has even been used for gauge "00" locomotive models. The wheels themselves should always be completely finished before the axles are made, so the operation of wheel turning will now be described.

Having selected all the wheel castings required, they should be carefully examined and gone over with a coarse file to get rid of any excessive sand or scale. Any uneven projections left from the moulding operation should also be removed by filing.

Plate 29 Gauge "1" wheel castings and a finished driving wheel, by Messrs. Bond's

The wheels are now set up in turn in the lathe, using the self-centring chuck, gripping them by the tread with the back of the wheel outwards. A suitable turning tool, as shown in Fig. 21, is now mounted cross-wise in the tool-holder and a cut is taken right across the back of the casting, leaving sufficient of the flange to allow machining of the front edge at a later operation. The wheel is now centred, using a stout centre-drill in the tailstock chuck and a drill is put right through the wheel with the lathe running at a high speed. In all the smaller gauges it is quite satisfactory to finish the wheels by the use of a reamer, in which case the drill used should be about 5 thou. under the finished diameter of the hole required. The reamer is now put through each wheel in turn, with the lathe running quite slowly; the reamer should be held in the tailstock chuck and the whole tailstock should be slid bodily forward, steadily in and steadily out, to obtain a good finish.

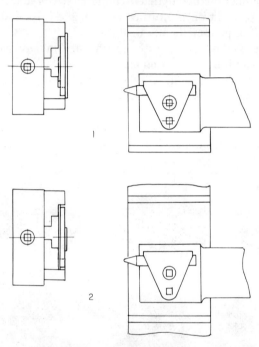

Fig. 21   Turning the wheels in a self-centring chuck. 1, Taking a cut across the back of the casting with the wheel gripped by its tread. 2, The wheel is gripped by its flange ready for machining the front face

It is sometimes said that by first chucking the wheel casting by its tread, back outwards, the operator is only setting the wheel to run true measured by the back, and it may be that the front of the wheel, much the more important side, may be out of truth. To obviate this, the wheel may first be chucked in the four-jaw chuck, by its flange, and set to run true by the spokes. A light cut is then taken over the tread, sufficient to "clean it up", after which the wheels may be chucked by the tread, back outwards, as described previously.

All the wheels should be treated in a similar way in turn, but before removing the partly turned casting from the chuck, a light cut should be taken over the flange to true this up and to provide a grip for the later operations.

In the next operation the wheel castings are reversed, so that the crank bosses are on the outside and the castings are held by the outside jaws of the chuck while the face of the rim and the crank boss are faced off in turn. It is not possible, of course, to machine the tread or the outer surface of the flange at this stage, as the lathe tool would foul the chuck jaws. The following

Fig. 22 The set-up for machining the tread and the outer surface of the flange. 1, The turning tool for machining the tread. 2, The lathe tool for chamfering the wheels. 3, The pointed lathe tool for cutting the groove between the rim of the wheel and the balance weight

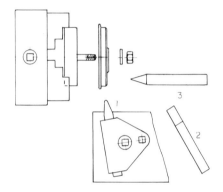

procedure should be adopted. A suitable iron casting, such as an old wheel or a small lathe chuck backplate casting, should be machined, faced off, centered and drilled and while still in position in the lathe a short length of round silver steel should be pressed in to this casting and then turned down to an exact sliding fit for the wheels to be dealt with. The end of this silver steel spindle should be threaded with any suitable BSF or BA thread and fitted with a nut and washer, so that the wheels can be put on in turn and nutted up tight against the back plate. It will now be appreciated that the backplate should be of such a diameter that when machining the wheel flanges to final size the lathe tool will just clear it.

The treads of the wheels and the outside face of their flanges are now machined and this should be done with a turning tool arranged as in Fig. 22, using the top-slide of the lathe with the saddle clamped to the bed. In this way the exact length of each cut can be measured by means of the top-slide graduations and by noting the reading on the cross-slide hand wheel, the diameter of all the wheels will come out exactly the same to quite fine limits.

It will be noted that the turning tool, as seen in the diagram, will leave the flange of the wheel of the correct size, except that it will be left square, leaving a sharp edge to be radiused off in a separate operation. This is sometimes done by means of a form tool, but it can also be done by applying a medium cut flat file to the flange with the lathe mandrel rotating. If the file is used for this operation, great care must be taken to avoid the file catching the chuck jaws as they rotate: a handle must, of course, be fitted to the file before use.

There are now two small further operations to be done to the wheels and these are the chamfering of the wheels, which is best done by a square ended lathe tool brought into the wheel at an angle of about 40 deg., as shown in Fig. 22, and the cutting of a small groove in between the rim of the wheel and the balance weight, by means of a pointed lathe tool, as shown in the drawing.

It will be appreciated that each of these operations in wheel turning should be carried out on all the wheels required for the model before the next operation is tackled.

Plate 30   Finished Gauge "0" wheels and axles, by Messrs. Bond's

Before completing the wheels there is one more important operation to be carried out and that is the drilling of the holes in the crank bosses to receive the crankpins. It is strongly advised that a small jig be made for this purpose and all that is required is a short length of flat bright mild steel, into which two holes are drilled and reamed at the exact spacing required, one hole being lightly countersunk to receive the drill, while to the other hole is fitted a short peg, consisting of a length of silver steel, which is an exact fit in the axle holes of the wheels being dealt with. See Fig. 23. Before drilling the wheel, the jig should be placed in position and located so that the crankpin hole comes nicely in the middle of the crank boss. The wheel should be set up on the table of the drilling machine and arranged on two lengths of square section steel, so as to allow clearance for the twist drill to pass right through the wheel. It is suggested that the drill used for drilling for crankpins should be a few thou. under the final diameter required, after which the holes can be finished with a hand reamer.

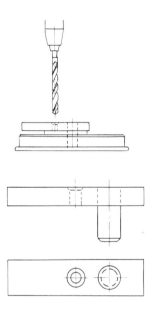

Fig. 23 Jig for drilling the holes in the crank bosses

Axles

Driving axles for gauge "0" and gauge "1" locomotives should either be made from silver steel or from ground mild steel. The gauge "0" driving axle may be $\frac{3}{16}$ in. or $\frac{1}{4}$ in. diameter, although the former size is the more common; driving axles for gauge "1" may be made $\frac{5}{16}$ in. diameter. Table 2 gives the standard dimensions for wheels and axles and also suggested dimensions for crankpins for gauges "0" and "1".

The three-jaw chuck can generally be used for axle turning, but if one or two of the jaws are out of truth it may still be possible to get the axle to run true by the simple expedient of inserting small pieces of thin paper between the offending chuck jaw or jaws and the axle material. Fortunate owners of collets of the right diameter will, of course, use these for axle turning in preference to the chuck.

If the cross-slide and top-slide of the lathe which is being used for axle turning have graduated dials, it is very easy to turn the ends of the axles to exact diameter required for a press fit and also to turn the wheel seat the

TABLE 2

Standard Dimensions

WHEELS

	A	B	D	E	P	R
Gauge "0" (Coarse) (32·0)	5·00	3·50	1·50	1·50	0·50	0·50
Gauge "0" (Fine) (32·0)	3·75	2·75	1·25	1·00	0·50	0·50
Gauge "1" (Coarse) (44·45)	6·00	4·50	2·00	1·50	0·50	0·50
Gauge "1" (Fine) (45·0)	5·00	4·00	2·00	1·00	0·50	0·50

LIMITS

Column A: $+0·005, -0 1$ and 1F
Column D: $-0·005, + 0$ in all gauges
Column E: $-0·005, + 0$ in gauges 0 and 1 and 1F

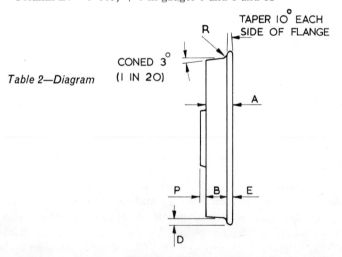

Table 2—Diagram

correct length to take the wheels, which of course will have been previously dealt with. It is advisable to turn a sample axle from a length of scrap steel of the same type as is to be used for the axles and try one of the wheels on for the correct fit. To make this operation easier after the axles have been turned to the correct press fit for the wheels, a very slight taper is put on the end, using a dead-smooth file, with the lathe running at a fair speed. This will enable the wheel to be "started", after which the dummy axle can be taken out of the chuck and clamped firmly in the vice. If the wheel is now gripped firmly in the hands and pushed home with a twisting motion, it should be possible to get the wheel halfway on if the wheel seat is the correct diameter. If the wheel can be pressed right home by this method, the axle wheel seat is slightly under size, so a suitable adjustment must be made to the lathe cross-slide.

Crank-axles

Those model locomotive builders who are particular as to appearance and are building a model of a prototype which has inside motion, may wish to fit a proper crank-axle. This is especially important in some inside cylinder engines, where the boiler is high pitched and where in the model the motor is fitted in the tender.

It is true that the making of a model crank-axle deters many builders from modelling an inside cylinder prototype, but for those who have a lathe and are prepared to tackle the job with care and in a logical sequence, it is not so difficult as might be imagined.

There are three principal methods of making crank-axles: (1) By brazing, (2) By adopting a built-up construction, all the components being press fitted together and (3) by turning from the solid. It might be mentioned that in the case of gauge "0" crank-axles having two throws at 90 deg. to one another, it is possible to reduce the amount of rough turning by cutting the crank-axle from flat bar and after partly machining, the crank-axle is heated to a bright red colour and the two sides are then twisted to the required angle. This method is shown in Fig. 24.

The main problem in the brazed type of crank-axle is the possibility of distortion after the brazing has been completed, while the press fit method is quite satisfactory but does call for some very accurate turning.

There is a method whereby the completed crank-axle can be maintained truly concentric throughout its length and this method will now be described. First, a completely plain axle is taken and the four webs are cut from flat steel of the required thickness and either drilled in a jig or alternately soft soldered together (all four), set up in the lathe and turned and bored as a single component. The hole in the webs must be turned to very slightly smaller internal diameter than the diameter of the axle itself (an interference fit of about ·00075 in. is satisfactory for gauge "0" and gauge "1" axles).

The two crankpins are now turned from silver steel of the finished diameter required to suit the inside connecting rod big-ends, the crankpins being

turned down at each end to a press fit in the webs (in this case the interference fit should be ·001 in.). The next operation is to press home each
pair of webs on its respective crankpins and to ensure that they are correctly
lined up while pressing the webs on, a length of round steel rod a close but

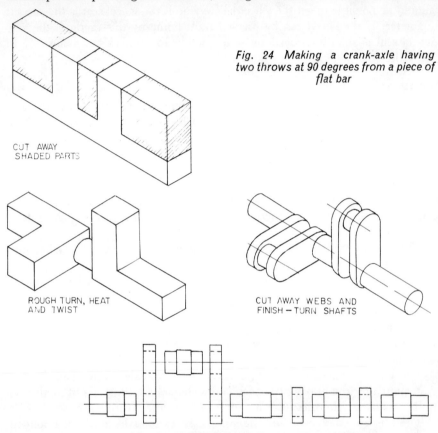

Fig. 24 Making a crank-axle having
two throws at 90 degrees from a piece of
flat bar

CUT AWAY
SHADED PARTS

ROUGH TURN, HEAT
AND TWIST

CUT AWAY WEBS AND
FINISH – TURN SHAFTS

Fig. 25 Making the crank-axle from four webs of flat steel
and a plain axle

not tight fit in the axle holes of the webs, is machined up and put through
the two webs while they are being pressed home. We now have to press the
two pairs of webs complete with their crankpins on to the axle without
distorting them and this is done in the following manner. A short length
of round steel of about twice the axle diameter is set up in the lathe, faced off
and drilled an easy fit for the axle. It is next reversed in the chuck and faced

off to an exact fit between the webs. It is sawn right across into two halves, all the burrs removed and it is then put in place between the webs, as shown in Fig. 25, ready for the pressing operation, which can be done in the lathe, unless the machine is a very light one, in which case the vice must be used.

The first pair of webs with this "split spacer" in position is now pressed home to its correct position on the axle when the "split spacer" is removed and the same operation is then carried out on the second pair of webs, care being taken to ensure that the two "throws" are at right angles to one another before pressing home the second pair of webs.

Crankpins

There has always been a good deal of controversy over the type of crankpin to be used for gauge "0" and gauge "1" model locomotives. While it is true that for "00" gauge models shouldered screws are quite satisfactory, these are not really suitable for the better class "0" gauge locomotives, while for gauge "1" models full-size practice can be followed quite closely.

The author therefore recommends that plain holes should be drilled in the driving and coupled wheels, finishing with a reamer where possible and the crankpins can then be turned to a press fit into the wheels.

It must also be remembered that many model locomotives are fitted with Walschaerts or Baker valve gear and these gears require a "return crank" fiitted rigidly on the end of the driving crankpin at a particular angle to the main crank. In such cases, therefore, it is most important that not only is the crankpin a good fit in the wheel, but it must be prevented from rotating, otherwise the timing of the valve gear will be upset.

Another point to watch is the leading crankpin on outside cylinder 4–6–0 and similar wheel arrangements, where the connecting rod works very closely to the crankpin of the leading coupled wheel. One solution to this is a rather thick washer, countersunk on the outside to take a countersunk screw fitting a hole drilled and tapped inside the crankpin.

Trailing coupled wheels normally carry a crankpin fitted with a hexagon nut and washer or sometimes a circular nut having two flats on it and in the larger scales a very small diameter hole can be drilled through this retaining collar or nut to take a split pin or even a miniature taper pin.

Coupling Rods

The ideal material for coupling and connecting rods in the smaller scales is undoubtedly nickel-silver. Mild steel or even stainless steel is occasionally used, but both these materials have some slight disadvantages. Mild steel, of course, rusts rather quickly, though it is actually the correct material as used on the prototype, stainless steel on the other hand, although completely free of rusting and corrosion problems, is a tough material to work, especially if fluting is required.

For "0" gauge coupling rods, metal about $\frac{1}{16}$ in. thick is suitable and

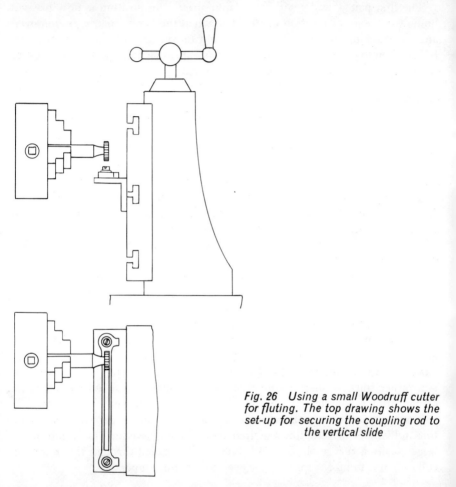

Fig. 26 Using a small Woodruff cutter for fluting. The top drawing shows the set-up for securing the coupling rod to the vertical slide

$\frac{3}{32}$ in. thickness is about right for connecting rods. For gauge "1", these thicknesses may be $\frac{3}{32}$ in. and $\frac{1}{8}$ in. respectively.

If the locomotive is fitted with sprung driving and coupled wheels, the coupling rod will need to be jointed and this joint is usually, but not always,

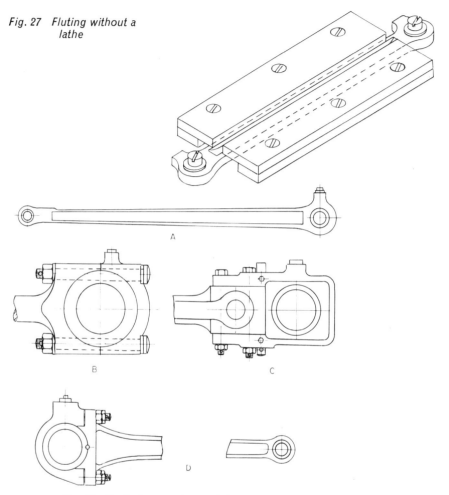

Fig. 27 Fluting without a lathe

Fig. 28   Typical connecting rod designs

placed just behind the driving wheel. The joint should be of the proper forked type. On some prototype locomotives, notably the old Great Eastern Railway 4–6–0's this joint was incorporated in the main bearing on the centre crankpin.

Whether the coupling or connecting rods are to be fluted or not, it is always advisable to mark out and drill the necessary holes for the crankpins first of all. It is generally sufficiently accurate to measure the spacing of the crankpins from the model itself by means of a pair of dividers, which are then transferred to the material.

There are several methods of fluting and if a lathe is available, one of the best is to use a small Woodruff cutter, held in the chuck while the coupling rod blank is mounted on a stout steel angle, which is in turn bolted to a vertical-slide, see Fig. 26. The advantage of this method is that the cutter will leave nice squared ends to the flute, whereas end-milling leaves a radiused end.

If no lathe is available, fluting can still be done if the following procedure is adopted. The coupling rod is held down to a suitable piece of steel bar clamped in the vice. Two lengths of thin steel strip, which should preferably be hardened, are then clamped down on top of the rod, so as to form a guide, as in Fig. 27. The flute is then formed by repeated strokes of a suitable scraper, which should be previously made to the width of the flute required or very slightly less, to allow for a working clearance.

Another method of forming a flute, which is quite feasible if the material is nickel-silver (but not for mild steel or stainless steel), is by cutting out the exact shape of the flute from thin sheet nickel-silver and soft soldering this to a strip of rather thicker metal of the same type.

The width of the material used for the making of both coupling and connecting rods should always be sufficiently wide to allow for the correct boss and the characteristic oil box on top. The coupling rods of most steam locomotives had circular bosses, but a few engines used a rectangular leading end boss, so as to provide for adjustment to the brasses. Connecting rods, on the other hand, were made in a wide variety of designs, some of which are shown in Fig. 28. While the bosses of the coupling rods for gauge "0" locomotives do not really need separate bushes, the fitting of phosphor-bronze bushes to gauge "1" rods is recommended, The intermediate joints and coupling rods do not need to be bushed, as the wear here is less than in the main bosses, but the pin used should be made in silver steel and as large a diameter as can be accommodated without spoiling the scale proportions of the rod.

CHAPTER SEVEN

Springing: Bogie and Pony Trucks

PROPER springing of the wheels of miniature locomotives in gauges "0" and "1" seems to have been rather neglected, possibly because the rigid wheelbase model is much easier to build. This is really rather unfortunate, because adhesion is greatly improved when all the driving and coupled wheels are in good contact with the rails all the time, regardless of irregularities on the track. Model locomotives with properly sprung driving and coupled wheels are less liable to de-railment than those with rigid wheelbases, because the wheels adapt themselves to parts of the track which are not so evenly laid as they should be. Thus when sprung locomotives and rolling stock are in use, the builder does not have to be quite so exact in his track building. This is an important consideration, for experience shows that most de-railments are due to low spots in the rails.

There is another, perhaps less important, aspect of springing and this is that the actual sounds of the locomotive wheels passing over the pointwork or over the gaps in the rails always seems much more realistic. Another good argument for springing all the wheels of the locomotive is that much better current collection is obtained. This applies particularly to two rail where, in most cases, the treads of the wheels are relied upon to pick up the current.

It should not be thought that proper springing for model locomotives is very difficult to install. It is certainly rather more work, especially if the driving axle is properly sprung in addition to the coupled axles (by driving axle in this context we mean the axle carrying the final driving gear wheel). This does, however, involve a special form of mounting for the motor, to allow it to float with the axle. A quite reasonable springing arrangement can be obtained by leaving the driving axle rigid and springing all the coupled axles.

The very simple springing arrangement which often serves quite successfully for gauge "00", is not really adequate for the larger gauges. It is far better to make up proper axleboxes, preferably in phosphor-bronze, though a good quality hard brass could also be used. It is not essential to fit proper hornplates to the frames, as even if the thickness of the frames is only $\frac{1}{16}$ in., the axleboxes will have sufficient bearing surface to allow for reasonable wear.

Although the springing may be in the form of flat phosphor-bronze sheet or strip bearing on the tops of the axleboxes, a much better plan in gauges "0" and "1" is to follow more closely the arrangement used in full size practice, that is to say to fit one or two pins to the underside of the axleboxes. A hornstay is then fitted across the gap in the frames and ordinary compression springs used underneath the stay with adjusting nuts on the end of the pin, which is threaded to take them. This scheme is shown in Fig. 29.

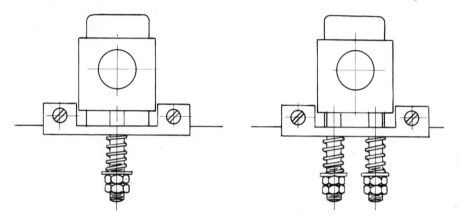

Fig. 29   Hornstay and spring pins under the axlebox

Very few enthusiasts have attempted a proper laminated spring in such a small size, but this form of springing should not be regarded as impossible. The usual difficulty is to obtain material which is flexible enough and yet has appreciable thickness, so as to give a reasonable scale appearance. The best material to use is Tufnol of a thickness a little under $\frac{1}{32}$ in. Experience has shown that a laminated spring made up of three or four leaves of thin Tufnol retains its resiliency almost indefinitely, especially if a single top leaf is used made of proper spring steel.

For large and heavy gauge "1" locomotives, a good laminated spring can be made up using three strips of Tufnol of 22 or 24 S.W.G. and with the top leaf made of spring steel of a thickness 30 to 34 S.W.G.

Coming now to the type of springing where the driving axle is sprung in a similar way to the coupled axles, a method has to be found of pivoting the motor on the end furthest away from the driving axle. Very often it will be found that a plate can be attached either to the back or the top of the magnet and a strip of thin hard metal, such as phosphor-bronze, is then attached to this and at the other end is fixed to a cross stretcher, which is pivoted directly in the frames of the model. See Fig. 30.

At the other end of the motor it will be necessary to make up a simple form of gearbox, so as to keep the two final gears in correct mesh. This again can generally be arranged without difficulty, either by the use of a gunmetal or brass casting or by building up the case of the gearbox by silver soldering together suitable pieces of brass sheet or strip and then soldering the whole to the overarm of the motor.

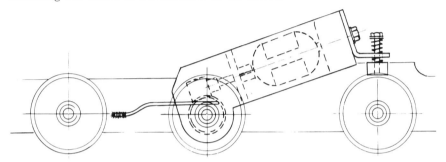

Fig. 30   Plate attached to the back of the magnet and fixed to a cross stretcher pivot directly between the frames

Bogies

The design of bogies for gauge "0" and gauge "1" locomotives is not generally given the attention that it deserves. While the bogie should always be made as heavy as possible, some form of springing should also be adopted.

The type of bogie where the pivot is arranged to the rear of the rear bogie axle (as often seen in tinplate models) should be avoided in a scale model. It is far better to follow more closely full size practice and a simplified type of the Adams' bogie is the one most suitable for the sizes of model locomotive under consideration. A central pin is used, attached to a cross stretcher fixed to the main frames of the locomotive and this pin passes through a sliding block which is free to move to each side of the bogie frame, so as to allow the bogie a reasonable amount of sideplay as the engine is traversing curved track.

The bogie pin is best made with both ends threaded so that the upper end can be screwed tightly into the cross stretcher; the plain centre part of the pin then passes through the sliding block while the thread on the lower end takes a retaining nut and washer. A light coil spring can then be inserted between the cross stretcher on the locomotive and the sliding block in the bogie.

Although seldom attempted, there seems no reason why a simple form of side control springing should not also be fitted, especially in gauge "1" locomotive work. This could take the form of two pieces of very light gauge metal, such as phosphor-bronze or spring steel, arranged so that they bear upon the outer ends of the sliding block, thus giving the whole bogie a self-centring action. Another method, rather closer to full size practice, is to fit two very small compression springs, one on each side of the sliding block, the outer ends of the springs bearing against the insides of the bogie frames, while the inner ends of the springs can be arranged in recesses in the sliding block, so as to keep them in place. See Fig. 31.

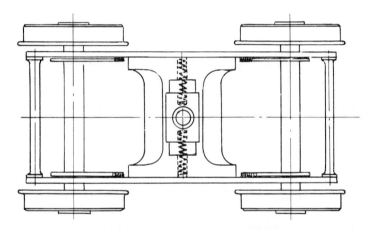

Fig. 31   Compression springs each side of the sliding block to provide side-control springing

Individual springing of the bogie wheels is not often attempted in these sizes, but here again there seems no reason why the more ambitious builder should not attempt it. For a gauge "0" locomotive bogie it would be possible to use brass strip about $\frac{3}{32}$ in. thick with oval slots cut in the frames to allow the bogie axles a little up and down movement. The bogie wheels are then fitted to their axles in the usual way and the springing is provided by thin gauge hard phosphor-bronze or stainless steel wire. The wire is attached to the back of the bogie frame and is allowed to bear directly on the tops of the axles. It might be thought that this method would cause too much friction, but if the wire is light enough and the axles are kept well lubricated, there should be no trouble on this account.

The construction of the frames for a locomotive bogie can follow that used for the main frames, although of course on a smaller scale. The shape of the side frames is marked out on one strip of metal of suitable thickness, this is then soft soldered to a second strip and the pair of frames drilled, sawn and filed to shape together, so that both the frames are an exact match. If the bogie frames are to be rigidly fixed to the bogie cross stretcher, the stretcher can be made of similar strip or sheet metal, as used for the side frames, though possibly a little thicker, but if the centre of the bogie is to be fitted with a sliding centrepiece for proper side control springing, it must be carefully marked out and cut to shape before the side frames are attached.

Plate 31 Underside of a Gauge "1" 4–4–0, showing simple bogie with individual springing

When assembling the bogie, it is a good plan to insert through the axle holes two lengths of silver steel of the same diameter as the bogie axles; this is so that it can quickly be seen whether the holes have been drilled accurately in line and whether the side frames have been accurately aligned one with another. Naturally, if the bogie wheels are to be individually sprung, as described previously, this cannot be done as the holes in the bogie side frames will be oval; but in this case the frames can be lined up quite accurately by keeping them held firmly down on a surface plate or some other flat surface while securing them to the cross stretcher.

As some of the older full size locomotives often had visible compensating bars on the outsides of the frames, these should be attached to the bogie frames before the wheels and axles are put in place. Such details as guard irons should also be fitted to the frames before assembly.

Some bogies, especially those fitted to certain classes of Great Western locomotives, had outside frames with springs and horns fitted to the outsides of the frames. This type of construction generally calls for castings and in some cases suitable scale die castings can be obtained from the Trade. However, if no suitable castings are available, it is probably best to attempt to build up the springs, horns and axleboxes in a similar manner to the full size locomotive. At least there is one advantage in the outside frame bogie, in that there is more space between the wheels for the fitting of the sliding block and the side control springing arrangement.

Finally, there is a special type of bogie which should be mentioned, that is the kind which was fitted to the well-known "King" class locomotives of the old Great Western Railway. In this bogie, as many readers will be aware, the leading bogie wheels were fitted with outside axleboxes, springs and horns, while the rear bogie wheels were supplied with the more usual inside axleboxes. The bogie frames, of course, had to be bent forwards and outwards to enable this to be done. In the full size locomotive, this unusual arrangement was adopted to ensure that the bogie wheels did not foul the cylinders, as the rear bogie wheels came rather close behind the outside cylinders, while the leading bogie wheels, which were rather close to the main frames of the engine, were given extra working clearance by "dishing" the frames immediately above and behind these wheels.

Pony Trucks

Little need be said about pony or two-wheeled trucks. The distance between the pony truck pivot and its axle centres may be safely scaled down from the full-size locomotive being modelled; a common error is to make this dimension too short.

Although all "full-size" pony trucks are fitted with some form of side-control, this is not really essential in the very small scales. If it is desired to provide some form of self-centring, a simple wire spring can be attached to the engine frame and its end located in a suitable lug in the centre of the pony truck.

CHAPTER EIGHT

Cylinders, Motion Work and Valve Gears

OUTSIDE cylinders for gauge "0" and "1" models can be made in three ways, cast in white metal, brass or gunmetal, cut and filed from solid brass bar, or built up from sheet metal. All three methods are quite satisfactory, although a brass or gunmetal casting is much to be preferred to white metal as this metal is very difficult to solder, its melting point being only slightly higher than most soft solders.

If a casting is used, it may have valve chest bosses and cylinder covers cast integral, in which case some careful four-jaw work in the lathe will be called for.

Lathe work will also be useful if a solid bar construction is preferred; in any case, the bosses and covers would have to be turned.

If a built-up construction is desired, the ends and bolting face can be filed to shape from $\frac{1}{16}$ in. or even $\frac{3}{32}$ in. thick brass sheet, while the walls should be made of quite thin nickel-silver, 10 to 15 thou. for gauge "0" and "1" respectively.

The usual procedure with built-up construction is to hold the ends square while the walls are wrapped around the ends and soldered to them. One method is to solder the ends to the bolting face first, as in Fig. 32; but a better method is to temporarily hold the ends at the right distance apart by means of two long screws or pieces of threaded rod, 8 or 6 BA, with nuts inside and outside, for which purpose the ends are first drilled on the centres of the valve spindles and piston rods, as in Fig. 33. The sheet metal forming the wall is then wrapped around and soldered to the end pieces. After removal of the screws or threaded rod, the cylinder covers and valve chest bosses are turned up and made a tight hand push fit in the end pieces; they should be provided with spigots to ensure correct alignment and, with the exception of the rear covers, which carry the slide bars, may be lightly soft soldered after final assembly.

The fixing of the slide bars will be dealt with later, but the rear covers may be made a very light press fit, or in gauge "1" work, bolted to the rear end of the cylinder block, using 12 BA bolts or screws.

As far as dummy bolt heads on front cylinder covers are concerned, it

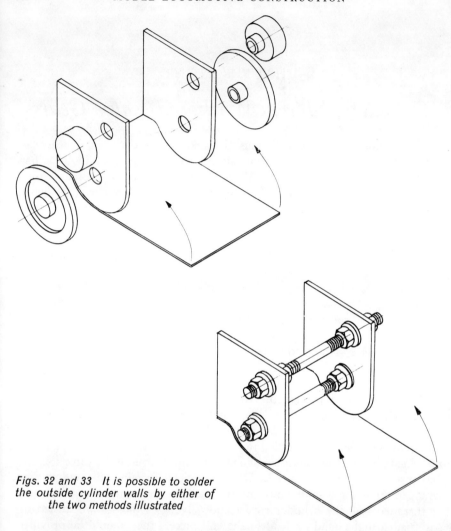

Figs. 32 and 33   It is possible to solder
the outside cylinder walls by either of
the two methods illustrated

should be remembered that, generally speaking, these are only exposed on earlier prototypes, and if they cannot be represented reasonably closely to scale, are best omitted altogether.

Do not forget the packing glands and stuffing boxes, which are quite easily represented if the rear covers are turnings. The valve spindle guides, which are often rather prominent on the more modern designs using Walschaerts valve gear, will need careful consideration according to their design.

Very often, they can be made from nickel-silver sheet or strip about $\frac{1}{32}$ in. thick, soft soldered to the rear valve chest bosses on either side of the valve crosshead, which is fitted with a pin working in the slots of the guides.

Cylinders are best fixed to the frames by two or three hexagon head screws, put through clearing holes in the frames, into tapped holes in the back of the cylinder. The use of hexagon head screws avoids the difficulty of trying to reach slotted head screws in the confined space between the frames. Even if the cylinder is of built-up construction, this is quite sound provided that the bolting face of the cylinder is not less than $\frac{1}{16}$ in. thick.

Cylinders should never be soldered to the frames; this is a poor method even in the smaller scales.

The small details, such as drain cocks and pressure release valves, which are generally found on locomotive cylinders, can be represented by wire and pins, or in the larger scales, by fine brass turnings. Attachment of these details is sometimes a bit of a problem, but if they can be threaded with a fine thread such as 12 or 14 BA, screwing them into the cylinder block makes a much neater job than trying to solder them. If the cylinders are solid, their rapid absorption of heat makes soft soldering very difficult.

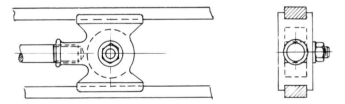

Fig. 34   Crosshead with two single slide bars

Crossheads

One of the most important components in the locomotive motion work is the crosshead. If the model is an inside cylinder one, we may not need to bother with crossheads at all, unless of course the builder wishes to fit a dummy inside motion. Crossheads are somewhat troublesome items to make, but before their construction is described, perhaps a few words on the various designs used in locomotive work will be of interest.

By far the commonest type is the crosshead which works between two single slide bars arranged one above the other. This is shown in Fig. 34. The sliding surfaces, generally known as the "slippers", are generally of equal length, but on some railways, notably on the old Great Northern railway, the upper "slipper" was made considerably longer than the lower, as in a locomotive which spends most of its time working in a forward direction, the wear takes place on the upper "slipper" and upper slide bar.

Another well-known type of crosshead is that using a single slide bar immediately above the piston rod and this type was common on the old Great Eastern Railway and on some other railways; it is shown in Fig. 35.

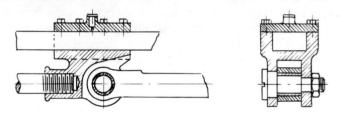

Fig. 35   Crosshead with single slide bar

Another well-known type is the three-bar crosshead, which was used extensively by Sir Nigel Gresley, and was also adopted for some of the British Railways standard locomotives.

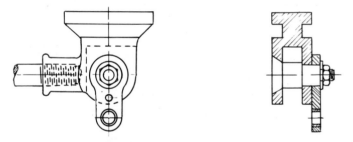

Fig. 36   Crosshead for three slide bars

The Southern Railway used rather an elaborate form of crosshead, which worked between four slide bars; this was used on the well-known "Schools" class and the "Lord Nelson" and "Pacific" types. A lesser known type of crosshead was one using four parallel slide bars quite close together, the crosshead consisting of two "slipper" blocks, each running in a pair of bars, with the small end of the connecting rod located between them. This design was used on some of his Compounds by Webb of the old London and North Western Railway.

The method of construction will vary according to the facilities available. If a lathe fitted with a vertical-slide is available, the straightforward two-bar crosshead can be cut from the solid without too much difficulty, but a better method is to build it up in three parts, consisting of the two "slippers" and the boss plus the central body. The best material is undoubtedly nickel-silver, though mild steel or even gunmetal may be considered.

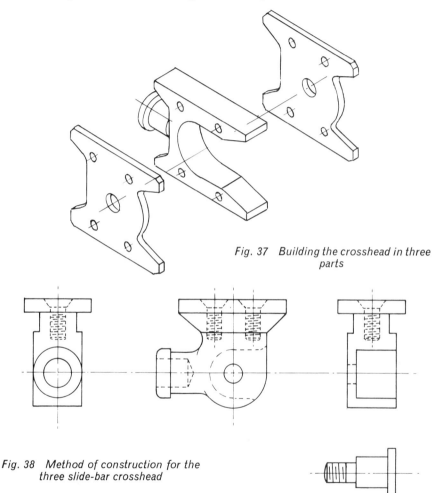

Fig. 37 Building the crosshead in three parts

Fig. 38 Method of construction for the three slide-bar crosshead

The single bar crosshead really calls for a simple slotting operation, to cut out the rectangular slot which is afterwards completed by soldering a flat strip of metal across the top. The type of crosshead which uses three or four slide bars is not at all easy to make in the smaller scales; again it calls for some rather fine milling work to ensure a good job. A suggested method of construction is shown in Fig. 38.

Slide Bars

Slide bars, whatever their design, are generally secured to bolting faces, machined on lugs on the back cylinder cover. Although in gauges "0" and "1", it is very difficult to adhere strictly to full-size practice, the slide bars should be firmly anchored to the cylinder cover, either by a small screw or by silver soldering. Slide bars that are merely butted up against the rear cover and soft soldered, never stand up for long and can be a source of continual trouble. Even in gauge "0" models, it is a good plan to make the rear cover removable from the cylinder, so that the slide bars can be properly fixed to it before final assembly.

Expansion Links

Perhaps the most important part of a locomotive valve gear is the expansion link, sometimes known as the curved link. In actual fact, in some valve gears the expansion link is straight, for instance in the well-known Allan valve gear.

The commonest types of valve gear found on steam locomotives were the Walschaerts and the Stephenson link. The former was most often found on the outside of the engine, though it was occasionally used as an inside valve gear, as on some of the Great Western Railway express engines. The Stephenson gear, on the other hand, is nearly always used between the frames, though towards the close of the L.M.S. railway as a separate concern, one of their standard 4–6–0's was fitted with the Stephenson valve gear arranged outside the frames operating the valves above the cylinders.

Other well-known locomotive valve gears are the Joy, which is generally used inside, and the Baker, which was at one time very popular in America, and which is only used outside the frames.

Towards the end of the steam era on the railways, various types of Poppet valve gear became popular and although this gear is difficult to model for working steam engines, it does not present many problems when used on electrically-driven models.

The author proposes to deal with the subject of valve gears in some detail, because he believes that many otherwise good electrically-driven models are spoilt by incorrect valve gear detail.

To take the Walschaerts valve gear first, almost all modern steam locomotives (i.e. post-1920) use cylinders with inside admission piston valves and the arrangement of this valve gear for this type of cylinder is shown in

Fig. 39, which also shows all the components of the gear with the names commonly given to them.

The most notable characteristic of the Walschaerts gear is that the movement given to the valve spindle is a combination of two separate motions, one produced by a return crank, which is mounted on the end of the main crankpin, while the other is produced by the crosshead and this second motion is 90 deg. out of phase with the first.

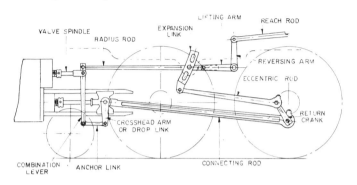

Fig. 39   The arrangement of the Walschaerts gear

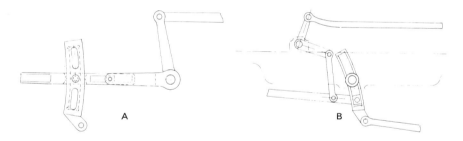

Fig. 40   Arrangement of the expansion link. A, L.N.E.R. and L.M.S. (later type). B, Southern Railway

It will be seen in the drawing that the valve rod can be moved up or down by the cab reversing gear, according to whether the locomotive is travelling forwards or backwards. When the valve gear is in its central position, it will be seen that the only motion given to the valve spindle is that provided by the main crosshead, and this is the mid-gear or the neutral position. When the valve rod is at its maximum lower or upper position, the engine is said to be in full forward gear or full backward gear respectively. When the valve rod is at some point between a full gear position and the neutral or mid gear position, the valve gear is said to be "linked up" (sometimes called notched up).

If the cylinders of the locomotive happen to be of the outside admission slide valve type (as was frequently used during the nineteenth century) the most notable difference found with the Walschaerts valve gear is that the rod is connected to the combination lever at a point below the valve spindle, while the return crank will be at a position 180 deg. from the position used with the piston valve type of valve gear. In actual fact, the return crankpin is arranged to follow the main crankpin by a nominal angle of 90 deg., whereas with the slide valve type of valve gear the return crankpin will be nominally 90 deg. ahead of the main crankpin.

Even if the model locomotive is an electrically-driven one, there is no reason why the outside valve gear should not be correct and function properly, even though the valve spindle is just a plain rod moving to and fro inside the cylinder.

Returning to the question of the expansion link, there are many designs of this component according to the layout of the rest of the gear. In most of the later L.N.E.R. and L.M.S. locomotives, the valve rod passed right through the expansion link and was raised and lowered by a lifting arm arranged behind the link. On some other railways, notably the Southern, the type of lifting gear preferred was one which was arranged ahead of the expansion link, as shown in Fig. 40.

Valve Gear Construction

Starting with the expansion link, if this is of the typical L.N.E.R. type, it should, strictly speaking, be made in three parts, the radius rod being slotted out, so that the central component of the link may work inside this slot. This is, of course, rather difficult an arrangement to make in the smaller scales and many builders may therefore prefer to adopt the typical L.M.S. type, where the link is basically of two plates with the radius rod passing through the middle. In this case the radius rod can be made from flat strip German silver (nickel-silver) or possibly in stainless steel. The longitudinal slot in the rear end of the radius rod can be cut by drilling a small hole at the rear end, inserting a fine metal-cutting fretsaw blade and making two cuts close together, according to the width of slot required. For finishing off such a slot the ideal tool is a fine flat needle file with the cutting teeth on one side only.

If the radius rod is of the type used with lifting gear in front of the expansion link, it will be quite an easy component to make, having no slots.

If it is desired to make the expansion link and the radius rod operate as in real practice, the curved slot in the link should be cut first in some nickel-silver plate and the outline of the link scribed out afterwards, as this will be found much easier than attempting to cut an exact slot in a link of finished outline. The slot may be cut as described for the radius rod, but it is a good plan to make the die-block first and use this as a gauge when finishing the curved slot in the link to size.

As regards fluting, this can be done as described in the previous chapter for coupling and connecting rods, a small Woodruff cutter being ideal for the purpose.

None of the other components of the Walschaerts valve gear should cause any real difficulty; those parts having forked ends should always be drilled first before the slot is cut. When the components are very short, as for instance in the case of the anchor link (sometimes known as the union link) it is a good plan to mark out the two links required on the ends of a piece of material longer than actually required, drill all the holes and then slot each end of the bar to produce the required fork. The links are then cut off the bar and the other ends of each link dealt with separately.

Stephenson's Link Valve Gear

As mentioned previously, the great majority of full-size locomotives fitted with the Stephenson link valve gear had this arranged between the frames; but if it is desired to fit an outside Stephenson gear, the main crankpin must be modified so as to carry the eccentrics or the return cranks necessary; two are required to each expansion link in this valve gear. The radius or valve rod is generally offset at its rear end and may be suspended by a simple lever, either at the back of the link or a short distance ahead of it. Some Continental locomotives fitted with the Stephenson gear outside the frames used eccentric sheaves and straps and in this case the eccentrics will have to be pinned to extensions of the main crankpin at the required angle.

Joy Valve Gear

The Joy valve gear was extensively used by some of the older British railways, such as the Great Eastern, North Eastern, Lancashire and Yorkshire and London and North Western railways. A few engines of the L.N.W.R. were fitted with an outside form of Joy gear and this gear was also used on some narrow gauge engines.

The Joy gear differs considerably from the valve gears previously described, in that no eccentrics or return cranks are employed. The drive is taken from a pin in the connecting rod, positioned approximately one third of its length from the small end. A link is attached to the connecting rod (this link is

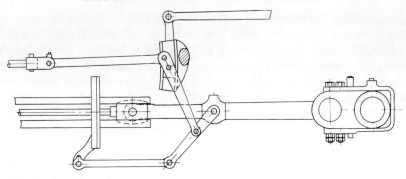

Fig. 41   The Joy valve gear

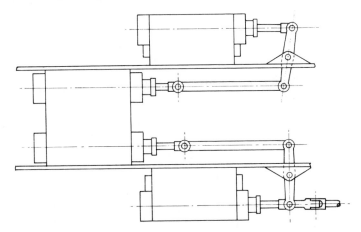

Fig. 42   Inside valves operated from outside valve gear

generally called the correcting link), the other end of this link being attached to an anchor link, which in turn is anchored to some part of the main frame. Further links, generally called the vibrating links, are attached to the correcting link, again approximately one third the length of the link from the pin in the connecting rod and the upper ends of the vibrating links carry two pins, one being attached to the radius rod and the other working in a die-block, which is able to slide up and down a series of curved slides, somewhat similar to the expansion links in the Stephenson valve gear, see Fig. 41.

Special Types of Valve Gear

The four cylinder Great Western locomotives were fitted with inside Walschaerts valve gear, the radius rods of which operated the valve spindles of the inside cylinders directly, and for the operation of the outside cylinders a form of rocking gear was provided, having a vertical pivot attached to the main frames. Thus, even the electrically-driven model, if it is to be realistic, should be provided with at least a simplified arrangement of the inside valve gear, so that a working rocking gear may be built up for operating the outside valve spindles.

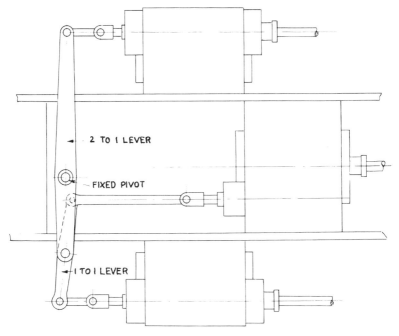

Fig. 43   The conjugated gear

Again, most of Gresley's express and mixed-traffic locomotives were fitted with a special form of conjugated gear, so that the outside valve spindles could operate the valve spindle of the inside cylinder. This conjugated gear is not at all difficult to make for the electrically-driven model, though it requires a simple form of frame stretcher. arranged between the frames, generally ahead of the cylinders, to carry the main vertical pivot, see Fig. 43.

CHAPTER NINE

Materials and Superstructures

AS has been explained in earlier chapters, the superstructure of the electrically-driven model locomotive is not much more than a shell and contains no working parts. The question of the best material to use for the superstructure is most important and will be considered first. At one time tinplate between 30 and 32 S.W.G. was the typical material used by amateurs for building the body of the locomotive, certainly in any gauge from "00" to gauge "1". Tinplate is very easy to solder and it takes paint extremely well; it is also very cheap. The main objection is that it rusts very quickly when sawn or filed and also at the edges if the paint should become chipped at any time. Another objection which has been made against the use of tinplate is that, being pricipally iron, it has some adverse effect on the permanent magnet of the electric motor. Nevertheless, if the problem of rusting can be overcome, tinplate remains a material which should not be dismissed lightly.

The next material to come into general use for small gauge locomotive work was brass, but although this material has, at first sight, certain advantages, it does not take paint very well and it is not quite so easy to solder as tinplate, as it allows the heat to dissipate away more quickly during soldering operations. The author suggests that brass should not be used for the flat parts of the model locomotive superstructure, though where it is decided to use a thin gauge seamless tube for the boiler barrel and/or the smokebox, it may not be possible to obtain this in the right diameter in any other material.

In the author's opinion, the ideal metal for the sheet metal parts of the model is nickel-silver, sometimes called German-silver. This material takes solder well, is completely rustless and does not corrode nearly so quickly as brass. Nickel-silver also holds the paint better than brass, though not so well as tinplate; it is reasonably easy to bend or to roll when used in boiler construction. It is important to remember that this metal can generally be obtained from metal merchants in three grades: hard, half-hard and soft. Although the metal can be annealed, this is a difficult operation and the amateur is advised not to attempt it.

For the turned parts, such as the chimney, dome, safety valves, buffers and so on, brass is quite satisfactory, though here again nickel-silver is the better metal where appropriate. In some cases, of course, especially on the older locomotives, the dome was left unpainted and polished, so that here it is essential to use brass. Some commercial fittings, such as domes and safety valves, are cast in a white-metal alloy; many of these are satisfactory, but they do not really compare with a proper turned fitting for the better class of model.

Where accessories can be fitted without the use of soldering, one of the light alloys, such as duralumin, may be considered.

For the running board angle or valance, brass angle or square bar or in some cases rectangular bar, is quite suitable, while parts such as buffer beams, which need to be a little heavier than the other flat parts of the superstructure, a rather heavier grade of nickel-silver or brass should be chosen. Finally, details such as handrails, water feed pipes and so on, are best made from some material which can be fairly easily soldered and yet are reasonably easy to bend. Stainless steel is not ideal, even for use as handrails on the smaller scale locomotive, as it is a difficult metal to solder. For any except the very smallest diameters, copper wire is ideal for water feed pipes and overflow pipes.

The Superstructure

The main components of the locomotive superstructure are as follows: the running plate, the valances (sometimes called the footplate angle or hanging bar), the buffer and drag beams, the splashers, the boiler/smokebox/firebox unit and the cab.

It should be explained here that the running plate in the model actually represents four quite separate plates in the full-size locomotive. In full-size practice, there is generally a separate running board alongside the main frames and the platform, known as the footplate, is really the floor of the cab.

In building the model locomotive, the running plate should form the foundation upon which all the other parts are assembled; it is thus most important that this be flat and true. If the running plate is a plain flat one, it should be a very easy matter to mark out and cut to size accurately; running plates with one or more bends in them will be dealt with later.

Choose as flat a piece of metal as possible for the running plate and commence by scribing a longitudinal centre-line right down its length and take all measurements from this. The sides and ends should be set out from the centre-line by means of a pair of dividers and the centre-lines of all driving wheels should be set out at right angles to the longitudinal centre-line, after which the cut-aways required for the wheels themselves should be marked out. The space necessary to clear the motor should also be marked out at this stage and after drilling four holes about $\frac{1}{8}$ in. diameter in the four corners of the part to be cut out, not too close to the scribed lines, cutting out may then be done with a metal fretsaw, finishing with files.

To assist following the scribed lines on the sheet metal it is a good plan not to clean this up until it is time to start soldering, or alternatively to give the metal a very fine coat of marking-out fluid or even quick-drying paint.

As soon as the motor cut out has been completed, the running plate should be tried in position on the chassis and the positions marked on it for the fixing holes, these being set off with the dividers to ensure that the holes come exactly in line with the cross-stretchers on the chassis. The sides of the running plate can be finished off dead straight by pressing a steel straight-edge or an engineer's square up against them, holding them up to the light to see if it penetrates at any point. The ends may then be finished off exactly at right-angles by using the square against the finished sides. After filing the cut surfaces with a smooth and then a dead-smooth file, a quick rub with some fine emery cloth will produce a nice finish.

Beginners should beware of the temptation to use a pair of tinsnips in work of this kind. Although the use of the snips may appear extremely quick, the metal is almost certain to be distorted and it will take much longer to flatten the material afterwards than it would if the metal fretsaw was used instead.

When the running plate has been cut out and finished to size, it should be examined to make sure that it is quite flat. If there is an even curve from end to end this will not do any harm, provided that it is slight, as it can be corrected when the valances are soldered on; but if there are any kinks they should be dealt with now before any attempt is made to erect the super-structure. Sometimes such kinks can be removed by careful bending with the fingers; but if this is not successful the job should be placed on a piece of heavy gauge flat steel and treated with some gentle taps from a light hammer.

The next operation is to attach the valances and the buffer and drag beams (if for a tank engine, the two buffer beams). For the valances, either brass or nickel-silver angle, square or oblong strip may be used, according to the depth of the valance required. Do not forget to allow for the thickness of the two beams when measuring the overall length required. Do not forget also that in almost all prototypes the running plate overhangs the buffer and drag beams very slightly.

Having cut the valances to size, lay them on the running plate with the buffer and drag beams pressed against them at each end, so that a check can be made that the dimensions are accurate. When soldering the valances to the running plate, the plate should be held down on a flat piece of hard wood with several drawing pins, which should be pushed well down so that the heads bed down against the metal and prevent it lifting, as in Fig. 44.

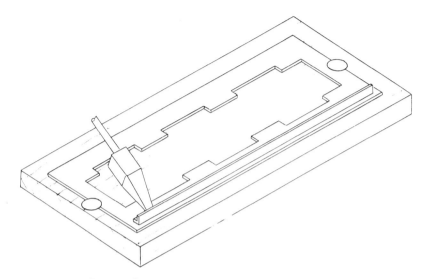

Fig. 44   Soldering the valances to the running plate

Now lay one of the valances on the running plate, set it back slightly from the edge, according to the design of the prototype, hold it in position with a small piece of wood and tack it with soft solder at about its mid point. Next check that the valance is in the right position and parallel to the edge of the running plate and if it is correct, work the soldering iron along the inside of the valance from the middle towards the ends, again holding it down with a piece of wood. Do not attempt to tack the valance at two points some distance apart and then fill in the intervening space; as the valance tends to expand more than the running plate, this method would almost certainly cause distortion. If the fillet of solder is somewhat on the heavy side, it should be cleaned up with scrapers before proceeding further.

After the valances have been properly fixed in position, the buffer and drag beams should be added. To simplify this job, obtain a true block of hard wood A, as in Fig. 45. Put the buffer beam up against the ends of the valances, push the block of wood up against the beam and then tack it down, checking that it is exactly at right angles. The bottom edge of this wooden block next to the buffer beam should be bevelled off just enough to clear the projecting edge of the running plate. The buffer or drag beam should, of course, be finished to shape and drilled and cut for buffers and couplings before being soldered into position. Now apply the soldering iron in the same way as described for the valances.

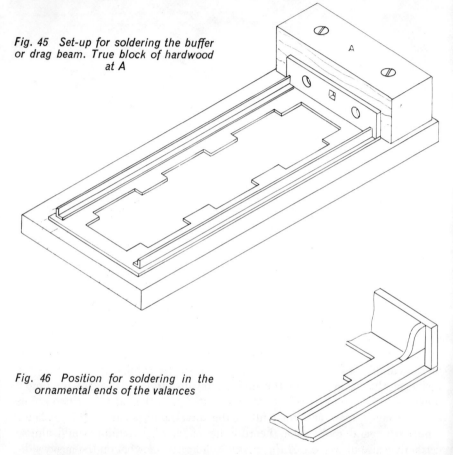

Fig. 45   Set-up for soldering the buffer or drag beam. True block of hardwood at A

Fig. 46   Position for soldering in the ornamental ends of the valances

It will now be necessary to add the curved ornamental ends of the valances up against the two beams and these little pieces can generally be cut out from the same material as was used for the valance and soldered in position, as shown in Fig. 46.

Remove the blocks of wood and all drawings pins and check that the structure, as so far completed, is quite true; if it is not, a little cautious bending with the fingers should put matters right.

Curved Running Plates

Most modern locomotives and nearly all express engines have running plates which are downswept at the front end in front of the cylinders and generally at the cab end also. Some running plates may have additional smaller bends in between the main bends, as in such engines as the old Great Northern "Atlantics", the L.N.E.R. "Pacifics" and some other types. Great care should be taken to ensure that all these bends are accurately made. The following method of assembly is recommended.

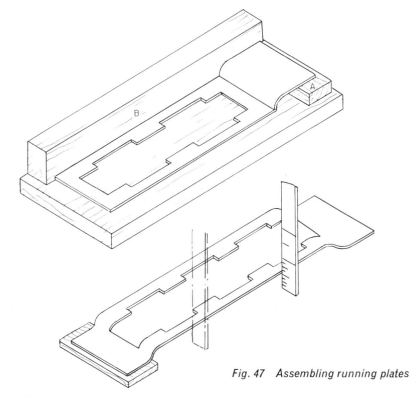

Fig. 47 Assembling running plates

As before, select a flat piece of hard wood to serve as an assembly bed; pin to this a block of wood A, the thickness of which is exactly the difference in height between the two levels of the running plate, as shown in Fig. 47. A third piece of wood B is pinned to the bed as shown to act as a guide to lining up the parts of the running plate. The main portion of the running plate is secured upside down with drawing pins or any other convenient means. The curved portion is then laid on block A, also upside down, and is again held in place with drawing pins. A good fillet of solder is now run right across the join on the inside, except near the edges where it would be in the way of the valances. If there is a downswept curve at the other end of the running plate, this can be fitted merely by reversing the plate in the jig, though if the bend is of a different level, the thickness of the block A must be adjusted to suit.

The middle or flat portions of the valances should be soldered to the running plate before the end parts, which should be carefully bent to conform with their curves.

Many enthusiasts have difficulty in obtaining the right radius when bending the short pieces of the running plate. One method which generally produces good results is to cut and trim up a length of metal about twice the length of the part required, but of the exact width, and put the bend in the middle of this, afterwards trimming off with the metal fretsaw. As to the exact radius produced, this is always a question of trial and error as it depends on the hardness or otherwise of the metal being used. If the metal used is in a soft or annealed state, it will be found that it will form a radius practically equal to that of the metal rod over which it is bent; but if the metal is in the "half hard" state (and this type of metal is recommended) the metal rod over which the material is bent should be of rather smaller radius than the required curve.

Fig. 48   Testing the shape of the running plate

The type of running plate in which the middle part above the driving wheels rises and falls in a continuous sweep or "reverse curve" is rather more difficult to produce. The centre part should be made in one piece and left a little longer than required at both ends for final trimming after the curves have been formed. The bending should always be carried out before the centre part is cut away to clear the motor, otherwise distortion is almost certain, the actual curves being formed by rolling over steel rods which can be held in the vice with sufficient overhang to allow the running plate to be applied.

When it is thought that the desired shape has been obtained, the work should be tested on a flat surface, as in Fig. 48. If there is a difference in height between the front and rear ends thin wood packing must be placed in position, as shown, to allow for this. The valance may be made in one piece unless there is an additional step down at the front or rear end, when it may be found easier to use a separate piece, as described earlier. In the case of gauge "1" locomotives, however, a method which can be used with success is to cut the whole of the valance out in one piece from fairly thick brass or nickel-silver sheet and this can be done by carefully marking out and checking against the shape of the running plate.

The Cab

The author usually prefers to assemble the cab before any of the other parts of the superstructure are dealt with. The two cab sides and the spectacle plate should therefore be cut out, trimmed to size, cleaned up and polished ready for assembly and a start should be made by obtaining another piece of wood A, Fig. 49, on which the running board with its valances and buffer

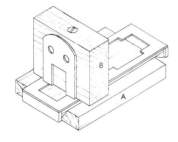

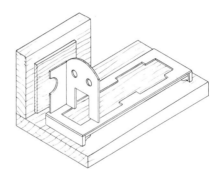

Fig. 49   The spectacle plate

Fig. 50   A projecting piece of wood goes over the running plate to support the cab side for soldering

beams is held down, only this time it is arranged the right way up. As the running board itself will now be lifted above the level of the bed by the depth of the valances, two strips of wood or cardboard should be put underneath the running plate just outside the motor cut-away, to prevent the running plate sagging in the middle. By means of another block of wood B, with previously squared sides, the spectacle plate may be held firmly in position square to the running plate; it is then soldered, the fillet of solder being formed on the side nearest the rear of the engine.

Now attach the two cab sides and this may be done as shown in Fig. 50, or alternatively they can be held in the fingers "freehand" or between two thin strips of wood. Once again, the soldering should be carried out on the inside of the cab. The roof should not be attached at this stage and in fact it is usually wise to leave this off until the model is nearly finished.

In some cases it will be found that the spectacle plate needs to be cut away in the centre to clear the motor; this should, of course, be done before assembly commences.

Wheel Splashers

In many cases the builder will find it easier to make and fit the wheel splashers, which protect the driving and coupled wheels, before mounting the boiler. In the smaller gauges we sometimes see the sides of the splashers bent up out of the material of the running plate itself; but this is not recommended in gauges "0" or "1", as it is difficult to get a sharp enough bend, apart from which it does not look right. It is best to make up the sides of the splashers first and solder them to the running plate, but to get them to stand upright and at the same time in their correct position in relation to the cutouts of the wheels, is rather a tricky operation.

Sometimes a length of strip wood laid along the outside will help to keep the splashers upright, while some constructors prefer to hold them in a small hand vice or in a miniature clamp and hold this in the left hand while the soldering iron is held in the right hand. Once the splasher has been tacked in place and in correct alignment, do not attempt to form a fillet right along its length, otherwise the solder is almost certain to melt and the splasher to fall away. It is better to form little fillets of solder, just sufficient to hold the splasher securely in position until the curved top plate has been attached. To make the top plate, cut out a length of metal of the desired thickness, cut and trim it to the exact width of the completed splasher and bend it by hand to match the curvature of the sides of the splasher. It will then be seen how much excess needs cutting off to fit properly in position. In some models the top of the splashers overlaps the sides by a small amount, so this should be allowed for when soldering up.

In bending the small parts, it is as well to remember that it is much easier if the bend follows the natural grain of the metal. The direction of the grain is easily determined by the slight curvature in the sheet metal when first purchased.

When all the splashers of the locomotive have been fitted in place and soldered, it will almost certainly be found that if the engine is of a large type the middle of the inner edge of the splashers will have to be removed to enable the boiler to seat properly at its correct height.

Boilers and Fireboxes

Boilers can be produced either from sheet metal or from tube. If tube is used, this should be as thin walled as possible (i.e. not thicker than $\frac{1}{32}$ in. even for a gauge "1" model). If thicker tube is used, the diffusion of heat through it will be so rapid that soldering small details to the boiler becomes very difficult. The ordinary electric soldering iron will be found unequal to the task, while if a large iron is used to produce sufficient heat to form a good joint, one joint is almost certain to come unstuck while another is being made.

If the barrel of the locomotive being modelled is a parallel one, the use of tube has a big advantage if the builder has the use of a lathe. It can be mounted between centres, or supported at the headstock end by the three-jaw chuck and by the tailstock (via a brass disc with a suitable centre in it) at the right-hand end. The boiler bands can then be formed on the barrel by taking a light cut from the surfaces between them. If this method of producing a boiler is adopted, great care must be taken to avoid chatter marks on the finished surface; the lathe should be run at a low speed, the turning tool should be similar to a rather narrow parting tool and accurately ground and stoned so as to cut freely.

In some cases, it is possible to machine the smokebox as well as the barrel from the one length of tube. Sometimes the diameter of the smokebox required may be appreciably more than the diameter of the barrel or even the diameter over the boiler bands; but even here the method can still be adopted by using rather thicker tube, turning down the barrel accordingly and very carefully boring out that part which forms the smokebox, so that the resulting wall thickness is not excessive.

If sheet metal is to be used for the boiler barrel, start by determining the circumference of the boiler. To do this, measure the diameter of the barrel as shown on the drawing and multiply this by 3·142. Lay this out on the sheet metal and mark off the overall length; the length of the smokebox is sometimes included, but this will depend on whether the smokebox is of appreciably larger diameter than the barrel itself.

Mark off from the rear end that part of the barrel which represents the firebox (assuming for the moment that this is a "round-top" type of boiler). The metal will now have to be cut up for a certain distance from each side with a fine metal fretsaw, to allow the firebox sides to turn outwards when the boiler is rolled, as in Fig. 51.

Also mark the positions for the boiler bands and the longitudinal centre-line along the top of the boiler, where the holes for the chimney, dome and safety valves will be located, but do not scribe too deeply or this will show up badly even after painting. Mark too the positions of any other holes that may be required, such as those for handrail knobs and check valves. To ensure that all the holes for the handrail knobs are exactly level, the boiler barrel should be erected on a surface plate, or something equally flat, so that the top is exactly horizontal; a scribing block may then be used to mark off the line of the handrail knobs.

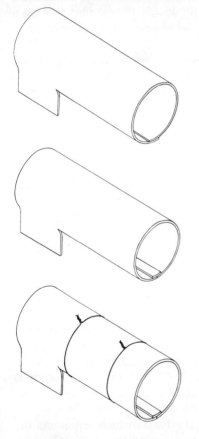

Fig. 51   Sheet metal boiler with an overlapping joint

Fig. 52   Sheet metal boiler with a cover "strap"

Fig. 53   Holding the boiler with loops of wire preparatory to soldering

In doing this, it should not be forgotten that an allowance must be made if the diameter of the smokebox is appreciably more than that of the barrel. This method is, of course, only applicable when the operation is carried out after the sheet metal has been rolled.

If the sheet metal being used for the boiler barrel is on the thin side, it is advisable to drill small diameter pilot holes at this stage and to enlarge

them to their final size after the boiler has been soldered up. The reason
for this is that the metal tends to bend more easily where it has been weakened
by the presence of a number of holes in line, so that it is difficult to keep
to the true cylindrical form when rolling up. If there is any doubt about
this, the holes may be merely centre-popped, leaving the drilling until
afterwards.

The rolling of boilers to obtain the true cylindrical shape is never easy,
but if the joint in the metal is arranged underneath, it is in this area that
most of the distortion is likely to occur and it will not be so noticeable after
other details, such as splashers, have been added. It is difficult to describe
a method of rolling a boiler which will turn out successfully in all cases, as
so much depends on the thickness and hardness of the metal. It will be
apparent that the softer the metal the easier it is to roll, but soft metals tend
to distort very much more easily than hard. To take an example, if the
normal commercial "half-hard" grade of nickel-silver is used, a considerable
amount of rolling will be required, using a round steel rod of a diameter a
little less than the boiler required. If available, a block of thick and not too
hard rubber laid on a firm flat surface will be found quite convenient on
which to roll the sheet metal.

Some builders prefer to use an overlapping joint at the bottom of the
boiler, in which case the width of metal used must have a sufficient allowance
for this, but in the larger scales and where the metal may be $\frac{1}{32}$ in. thick
or even more, better results are often obtained by butting the two ends of
the sheet metal together and using a cover "strap" as in Fig. 52.

It is a mistake for the unskilled amateur to try to anneal either brass or
nickel-silver. This is a skilled process, which is best left to the metal manufacturers. When the boiler has been brought closely to the final shape and
diameter required, it may be held temporarily in position while the soldering
of the joint is carried out by means of loops of soft wire, as shown in Fig. 53.
Such loops allow the slight increase or decrease of diameter if the sheet
metal has been lapped together. If these wire loops are made of slightly
rusty soft iron wire, the solder will not adhere to them. Now solder along
the seam, using sufficient solder to conceal the joint; do not worry unduly
if the boiler appears slightly oval at this stage, as this may be dealt with after
the soldering has been completed.

Having successfully shaped the boiler and cleaned up the soldered joint,
that part of it which must be removed to clear the motor should be taken in
hand. The width of the motor should be taken by means of a pair of dividers,
allowing a safety margin on each side, and the measurement can then be
transferred to the boiler. After bending the firebox sides, as required by the

design of the model, the unwanted part of the barrel can be removed with a metal fretsaw.

Taper boiler barrels

So far, we have only dealt with parallel boiler barrels. Taper barrels are naturally a little more difficult to deal with. An examination of full-size practice shows that the great majority of locomotives with taper barrel boilers had the barrel arranged with the bottom horizontal, so that all the taper was on the top. In the case of some of the smaller engines, the taper

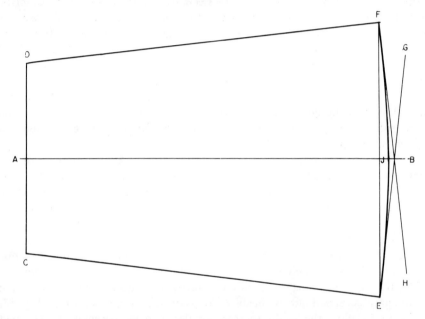

Fig. 54 Diagram to aid calculations for finding the correct proportions to construct the taper boiler barrel. The lettering is explained in text

was quite steep, and in certain examples, the barrel was made partly parallel and partly tapered. In the latter case, of course, the model barrel would have to be made in two quite distinct pieces, joined together.

Although some model locomotive builders go to great lengths to lay out a taper barrel in the flat to strict mathematical dimensions, this is not at all necessary in the smaller scales. The author has always found it easier to lay out a normal "true cone", roll the barrel up in the usual way, allowing some excess "tube" at both ends, then lay the coned barrel on something flat and mark off the ends at right-angles to the base, finishing by cutting and filing the ends as required.

To lay the material out in the flat, by the above method, the required diameters at the small end and the large end are first taken off the drawing, and the circumferences at each point worked out in the usual way (Circumference $= 2 \pi r$, or πD). A longitudinal centre line AB in Fig. 54 is now drawn on the sheet metal and the two circumferences CD, EF, set off at 90 deg. at a distance apart equal to the required length of barrel, plus a small allowance for trimming—say $\frac{1}{16}$ in. per inch of barrel diameter.

Lines are now drawn from C to E and from D to F. Now draw lines EG and FH at 90 deg. to CE and DF respectively as shown in Fig. 54. The required edge of material is then given *approximately* by the curved line EJF which can be drawn in quite easily with a suitable trammel. If a little extra length has been allowed at the small end of the cone, there is really no need to draw a similar curved line at this end, though if drawn, it will of course be curved towards the right-hand side of the drawing. Now CEJFDA is the shape of the metal required, though no allowance has been made for overlap. If overlap is required, this can be set off from CE in the drawing, being nothing more than a rectangle based on CE of the width required. (About $\frac{1}{8}$ in. for gauge "0", or $\frac{3}{16}$ in. for gauge "1").

Although the above method can be used directly on the metal to be rolled, it is not a bad plan to make up first a dummy barrel in thin cardboard, the longitudinal edges being fixed together with a quick-setting glue, when any corrections can be made, and the cardboard opened out again before the glue has finally set.

There are two other methods of making taper boiler barrels. One is to use a length of brass tube of a diameter equal to the diameter of the required taper barrel at the larger end. A vee-shaped piece is then cut out from one end, and the tube annealed and squeezed in until the cut edges meet. The other method is to use a length of tube of a diameter equal to that of the smaller end. This is slit longitudinally with a fine saw and is then forced open by hand until the firebox end is at the required approximate diameter. The cut *should not be made right through*, but about $\frac{1}{8}$ in. should be left to keep the tube together. A vee-shaped piece of tube is now cut out and inserted in the gap, and then soldered in place. At a later stage, it will probably be necessary to cut away a fair amount of this strip, to clear the mechanism, but once the smokebox and firebox have been completed and attached to the barrel, this will not matter. In any case, a narrow strip of metal could be soldered round the inside of the barrel to check any tendency to distortion.

Fireboxes

The firebox of a "round-top" boiler is no problem in an electrically-driven model, as it forms a part of the "barrel", but Belpaire fireboxes need special treatment. Perhaps the best way, in gauges "0" and "1" is to make a proper hardwood former. This should be made in three pieces with

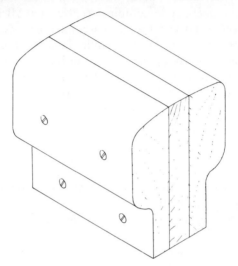

Fig. 55   The hardwood former used for making Gauge "0" and Gauge "1" fireboxes

a movable middle piece as shown in Fig. 55. The former is made as follows: mark off on the wood from the locomotive drawing the front elevation of the firebox, less the thickness of the metal to be used all around, and at the other end, the rear elevation of the firebox—remembering that in many engines, the firebox is both lower and narrower at the rear end. The piece of wood is now cut down the middle, the two parts being reduced on the cut surfaces so that a loose middle piece from $\frac{3}{8}$ to $\frac{3}{4}$ in. thick (according to the size of the firebox) can be inserted in the middle, as shown. The three pieces are then screwed together and the shape produced by planing, chiselling, and/or filing, not forgetting the rounded corners and possibly a sloping backhead, according to the propotype.

The firebox can now be formed around the wood former, the metal used being either soft brass or soft nickel-silver. After the top has been formed, the former is removed and the screws taken out, after which the former is replaced for forming the sides and waists. The former can now be removed by slipping out the centre piece, when the two outer parts of the former will fall out.

To form the front of the firebox, using a separate piece of metal, the former is re-assembled. If the firebox is one with well rounded corners, copper can be considered for the front only. The backhead is generally just a flat piece of metal, to which the cab fittings, if required, can be attached. When all parts have been formed, they are soldered together, and a strip of metal should be soldered inside where the joints occur between the sides and the front and back, to strengthen the whole.

Another method of making fireboxes is to mark off on the sheet metal the top and sides of the firebox, allowing sufficient for the radii of the corners, but ignoring the corners of the front and back plates. The plate is then bent by hand, by trial and error. A flat plate is then soldered to the bent-up sheet to form the front, and if the corners are well-radiused, as in most ex-Great Western and Stanier L.M.S. engines, a piece of thick square brass, say $\frac{3}{16}$in. for gauge "1", is bent up and soldered on the inside, between the front plate and top and sides. The corners are then filed away until the desired shape is produced.

Locomotives such as the Fowler types on the ex-L.M.S. and Midland Railway had fireboxes with rather sharp front corners, and the wrapper (or at least the outer cleading) was not waisted in, but came straight down to running board level, so that these are much easier to model, but the fireboxes

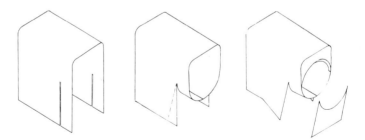

Fig. 56   Method for making L.M.S. "Pacific" fireboxes

of the ex-L.M.S. "Pacifics" present special problems, as they are really a combination of the Belpaire and Wootten types.

One method of making the L.M.S. "Pacific" type firebox is shown in Fig. 56. The sides are cut up as shown, with a fine fretsaw, so as to form the combustion chamber, the ends of the pieces bent up to form this being soldered together, after which the now open front of the firebox is filled in by a "throatplate". The remainder of the sides of the firebox are next cut and filed to shape, and finally a thick piece of brass is soldered in the top front corner of the firebox, which is filed as required to produce the rounded top edge and corners.

The Smokebox

If the boiler barrel has been rolled up from sheet metal, as described earlier, the smokebox may be formed by simply wrapping a further strip of sheet metal around the boiler, with the ends bent out to stand on the running plate of the locomotive. If the smokebox of the prototype is of the drum variety, a separate saddle will be required, so that in this case, the strip of metal will be wrapped right around the boiler and will be soldered along the bottom in a similar way to the barrel.

A hole will be required in the smokebox wrapper for a screw to hold the chimney down, unless this is to be fitted by means of small bolts around the base, as in full-size practice (very difficult in such a small scale) and this must correspond with that in the boiler; other holes will almost certainly be required for handrail knobs and possibly other fittings.

The smokebox front will next be required and there are several ways of making this. It may be machined in one piece with the smokebox door, or a separate thin piece of sheet metal may be used, to which the turned smokebox door is attached by means of the usual centre handle. The disadvantage of making the front in one piece is that this necessitates rather a heavy piece of metal which may cause difficulty when soldering.

The next operation will be to solder the boiler and smokebox unit to the running plate, Having located it carefully, it can be tacked in place with the soldering iron at one point only at some convenient place at the front end. The whole job should now be dealt with carefully, to make sure that the boiler unit is exactly in the middle of the running plate, that it is parallel with the edges and that the top of the boiler is quite horizontal. This latter requirement can be quickly checked (assuming that the model has a parallel barrel) with the scribing block, as shown in Fig. 57. If all is well, the soldering can now be completed, working the soldering iron inside as far as possible, in order to avoid the need for too much cleaning up. This, of course, is seldom possible at the front end. Do not forget to run a thin fillet of solder all around the rear end of the firebox where this butts up against the spectacle plate.

Now try the complete assembly in position on the chassis, to make sure that it clears the motor. If it is found to be fouling the motor at any point so that the superstructure does not lie evenly, this must be put right before proceeding further. It is, of course, most important to make quite sure that no parts of the insulated side of the brush gear can short-circuit against the superstructure; it is, in fact, advisable to give the model a preliminary trial on the track at this stage. Should the model not run satisfactorily and there is no sign of electrical "shorting", it may be that one of the wheels is fouling the running plate or one of the splashers.

When dealing with models of the earlier prototypes, it is often found that the smokebox is required to be considerably larger in diameter than the boiler barrel. In such cases, it is not easy to build up the necessary thickness simply by wrapping thin sheet metal around the barrel, as this would make soldering extremely difficult. One solution is to make up the smokebox as an entirely separate unit, using a single wrapper of sheet metal with identical front and back plates. Any beading on the joint can then be formed from wire of suitable gauge.

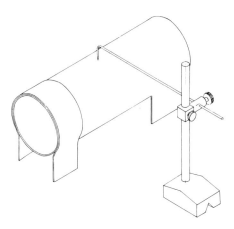

Fig. 57 Checking that the boiler top is horizontal with the aid of a scribing block

Boiler Bands

Although many builders fit boiler bands after the boiler has been mounted on the running plate, the author prefers to fit these before assembly, when access to the underside of the boiler is so much easier. It is very difficult to cut thin metal strip of the correct width for scale boiler bands, though it may be possible to obtain strip of the right dimensions from the Trade. Many otherwise good models are spoilt by boiler bands of too heavy a section, so that some means of cutting them from the sheet metal may have to be attempted. After the metal is cut with tinsnips it will be found that when laid out flat it is distorted into a curve. This may be remedied by first cutting the stock sheet metal on a curve in the opposite direction to the distortion.

Plate 32   Chimneys and domes (brass pressings) for Gauge
"0", by Messrs. Bond's

The boiler band is best cut about an inch longer than the circumference of the boiler, so as to provide a "handle" by which to hold it while soldering; the surplus may be cut off later. Another method is to cut the boiler band to the exact length required and start the soldering in the middle of the band on the top surface of the boiler. The band is then pulled round by the fingers or by tweezers as the soldering proceeds. It is a good plan to lightly tin the boiler band on the inside before commencing operations and the soldering iron should be kept in contact long enough to ensure that the tinning melts and forms a proper joint.

It is advisable to clean up all the boiler bands before any fittings, such as handrails or blower pipes, are added, as these would be a serious obstruction to the use of files, scrapers or emery cloth.

The soldering of boiler bands will be found much easier where the barrel has been made from thin sheet metal, as when brass tube is used the thickness of the metal causes the heat to be diffused very quickly, making it difficult to get the solder to flow freely when attaching the boiler bands.

Boiler Mountings

Many locomotive builders buy their turned boiler mountings, such as the chimney, dome, safety valves, etc., from the Trade; but if a lathe is available they are not very difficult to make, provided that the right methods are employed.

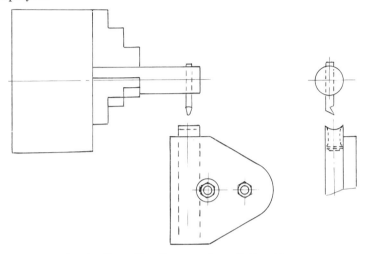

Fig. 58 Fly-cutting the curve beneath the chimney

As it is most important that the correct shapes are produced, especially those of the chimney and dome, great care should be taken over this. A full-size drawing is essential and it is advisable to have one or two photographs of the prototype for handy reference while working on the lathe.

The curved surface underneath the chimney and dome to match the radius of the smokebox or boiler barrel is often a cause of trouble to the beginner. While in the smaller gauges it is possible to obtain a reasonable fit by pressing the partly made boiler mounting (the metal having been previously annealed) on to a piece of round steel rod or tube of the same diameter as the boiler, this method is not recommended for gauge "0" or larger. A much better method is to fly-cut the underside. Some brass rod is obtained of a diameter slightly greater than the diameter of the base of the chimney being turned, and before anything else is done, one end of this is faced off in the lathe. This end is then fly-cut, using the cutter held in the lathe chuck, with the stock metal clamped under the tool-holder at right-angles to the lathe mandrel, as shown in Fig. 58. The stock metal must, of course, be clamped exactly at centre height and the fly-cutter must have its cutting bit set out to describe a diameter equal to that of the smokebox or boiler barrel, as the case may be.

The lathe is now set in motion, using a medium speed, and the saddle traversed across the cutter, using a light cut. When cutting has been completed, leaving a nice curved surface matching that of the smokebox, the stock metal is set up in the self-centring chuck and turning tools of various shapes are mounted in the tool-holder to turn the outside shape of the chimney or dome. The most useful shapes of lathe tools are as shown in Fig. 59.

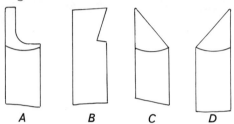

A   A type of parting tool
B   Side view of A
C   A useful R.H. tool
D   A similar L.H. tool

A   B   C   D

Fig. 59   Useful lathe tools

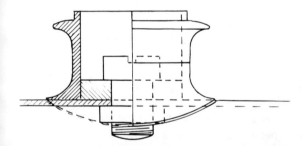

Fig. 60   Collar for mounting the chimney

It will soon become apparent that however carefully and accurately the turning is carried out, there is a portion of metal on each side of the "skirt" of the chimney which cannot be removed by normal turning methods. These parts may be filed away by hand if care is taken and this operation will be described later.

If it is a chimney that is being dealt with, the stock should now be drilled, using twist drills in the tailstock chuck and it will be found that the bore of the chimney must be made rather less than the strictly scale dimension, otherwise the metal left around the centre part of the chimney will be too thin for safety. The chimney may now be parted off and the final operations on the base started.

The principal difference when turning a dome, is that as much as possible of the curved top surface is machined in the lathe, finishing with fine files and various grades of emery cloth and the last stage may be tackled by carefully using a parting tool, so as to leave a little "pip" on the top of the dome. Although most model locomotive builders appear to solder their domes to the boiler, a much better method is to drill and tap the dome from the underside (which must, of course, be done before parting off) as this enables the dome to be screwed down to the boiler by a screw put through from the inside, which has the great advantage that the dome may be removed for polishing or repainting at any time.

Returning to the problems encountered in chimney turning, the method of drilling the chimney right through may seem to be faulty in one respect: that it cannot be bolted down to the smokebox until a small collar has been fitted inside it and soldered near to the bottom, as seen in Fig. 60. But this slight disadvantage is outweighed by the fact that a chimney made in this way can be mounted on a plain round mandrel for finishing operations, for painting or for polishing.

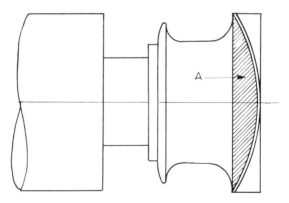

Fig. 61   Filing the chimney base

When satisfied with the outside shape of the chimney or dome, work will now commence on removing the two portions of metal, shown at A in Fig. 61, and this is best done by temporarily screwing the boiler mounting to any suitable piece of rod or tube of the correct diameter and gradually and very carefully filing away the unwanted metal. It is best to use a fairly coarse half-round needle file for most of this work and the filing should always be done on the right-hand side of the chimney, so that if the file should slip it will tend to slide away from the body of the chimney towards the edge, thus avoiding scoring the previously finished surface. In fact, it is

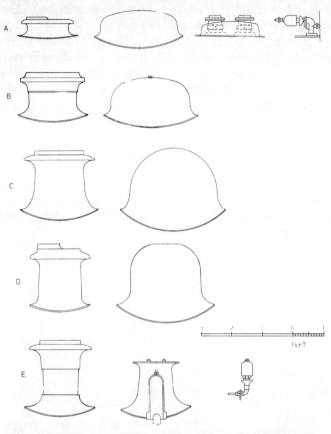

Fig. 62 Various types of boiler mountings: A, "Royal Scot"; B, typical L.N.E.R. types; C, typical Great Central shapes; D, Midland Railway mountings; E, G.W.R. "Prairie" tank engine mountings

not a bad plan to temporarily wrap around the centre part of the chimney or dome a piece of very thin sheet metal, to protect it against possible damage during filing.

CHAPTER TEN

Tank Engines

IN some ways, tank locomotives are easier to build than tender engines. Certainly they have more space in their superstructure for the accommodation of the power unit.

Construction of tank engines follows generally that advised for tender engines, though in certain cases it may be found more convenient to assemble the tank unit on the footplate before the cab is fitted. Much depends on the design of the prototype. After the tanks have been soldered up and fitted to the footplate, the cab front or spectacle plate is fitted, followed by the cab sides, except in designs where the cab sides are more conveniently made in one piece with the tank sides.

In the smallest gauges, the fronts of the side tanks are often made in one piece with the sides of the tanks, but this method is not recommended in gauges "0" and "1", unless the corner is a rounded one, when some careful bending will be called for.

In many tank engine designs, the tops of the side tanks are set a little lower than the top edges of the sides, and to make a good job of this, lengths of fine angle or square material should be soldered on the backs of the side tanks before assembly. It is also a good plan to solder a further strip of material diagonally across the back, to prevent distortion during assembly and to give greater strength all round. When fitting the sides of the tanks to the footplate, it is often necessary to do the soldering from the outside, so that the amount of solder used should be kept to a minimum, especially if rivet detail has been fitted to the tanks.

A point worth noting in models of most side tank engines is that a great deal of the boiler can be cut away below the level of the tanks, giving greater space for the motor, and making it easier to fit extra weights to the model, for better adhesion.

In some tank locomotives, notably many of the old Scottish prototypes, the cabs were considerable narrower than the side tanks. In this case, the rear ends of the side tanks will be completed by sheet metal soldered right across the model, with the usual cutaway for the motor, cab sides being added later and a short spectacle plate on top of the tanks. In some cases, the rear ends of the tanks will be completed as in real practice, with just a narrow strip, the cab sides butting up against this. In others, the cab sides will be short pieces extending down only as far as the tank tops, to which they are soldered.

Many of the old Great Western tank locomotives had side tanks which were flush with both the cab sides and the bunker sides, so the way to go about these is to cut the whole from one piece of sheet metal, the only connection between the front and rear parts being a narrow strip over the doorway, as at A in Fig. 63. But the weakness at this point can be overcome

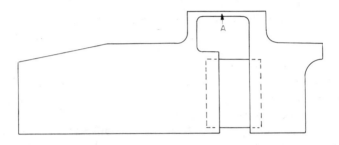

Fig. 63   Typical side tank and cab ride: G.W.R.

by soldering a cab "door" of the same gauge of metal as used for the cab sides on the back, across the opening, thus joining the front and rear sections firmly together. In some engines, this door was set back from the sides, and this can be represented by first soldering two strips of metal of suitable thickness along the two edges of the opening, on the back, then laying the door proper across the back of these strips.

Saddle Tanks

Saddle tanks make interesting prototypes, some were built by the old main line companies, but a much greater variety by the small "private" builders. The shape of the actual water tanks was often quite complex. At first sight, it might be thought that in front elevation, the shape was formed from three radii—two small radii along the bottom edges on each side of the engine and one much larger radius over the top of the boiler. But in many designs, this shape is formed from five or even seven different radii.

In some saddle tanks, the situation is complicated by the fact that the tanks are not continued right up to the spectacle plate, but a small gap is left in front of the cab, where safety valves may be mounted over the rear of the firebox. In any case, when the locomotive is being assembled, it is generally prudent to mount the boiler on the footplate in the usual way, then make and fit the saddle tank to it before tackling the cab.

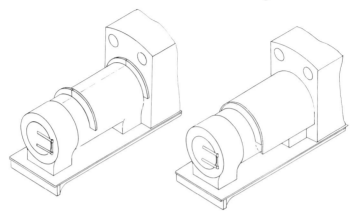

Fig. 64 The saddle tank, showing the front and back pieces with scribed lines (left), and the top-and-side piece illustrating the angle to which the bottom edges must be radiused over a piece of hardwood clamped in the vice

Start by cutting out and filing to shape the front and back pieces, which will be identical. The sheet metal for these parts may, with advantage, be made a little thicker than that normally used for model locomotive bodywork. Very lightly scribe a vertical line on each of these pieces exactly in the middle, and a longitudinal centre line should also be lightly scribed along the top of the boiler, so that the front and back pieces of the tank can be quickly lined up with the boiler. Before attaching these pieces to the boiler, the top and sides (in one piece of course) of the tank should be bent up; it should be made of fairly thin sheet metal, to facilitate correct bending.

To obtain the length of material required for this part of the tank, bend up a piece of thick paper and trim it to shape, then open it out again and measure it. A small margin of additional metal should be allowed for possible error, say $1\frac{1}{2}$ mm. for a gauge "0" model. The tank top is now bent to shape in much the same way as for a boiler. Start in the middle and work outwards, offering the front piece up repeatedly as bending proceeds, so that any adjustment necessary can be made early on. The much smaller radii at the bottom corners of the tank can be formed by bending the metal over a piece of hardwood of suitable shape in the vice. Care must be taken with these lower radii, as being much smaller than the remainder of the tank, little or no adjustment is possible after bending has been completed.

Having successfully shaped the top of the saddle tank, start assembly by soldering the front and back pieces to the boiler, lining them up by means of the scribed lines referred to previously, and by a piece of hardwood to ensure that they are at right-angles to the top of the boiler. Tack them lightly with solder at first, then check for truth, finally completing the soldering on the inside, where the joint will not show. The top part is now taken in hand, drilled for any details required, dome, etc, and these should preferably be screwed on before the top is soldered in position.

The attachment of the top of the saddle tank to the front and rear pieces should be started at the mid point of the front piece, the sheet being tacked only at first, then the metal can be held firmly against the front piece by a piece of hardwood, while soldering proceeds, tacking a point on each side at approximately the same distance from the middle, to avoid distortion. The rear can then be tackled, after which the whole tank is cleaned up, ready for the attachment of the spectacle plate and the cab sides.

Pannier Tanks

Pannier tank locomotives were the speciality of the former Great Western Railway, which had quite a variety of different classes with driving and coupled wheels from 4 ft. $1\frac{1}{2}$ in. diameter to 5 ft. 2 in. They make exceptionally interesting models.

The most notable point about pannier tanks is that almost the whole of the boiler and smokebox is hidden by the water tanks. For this reason, models made in the smaller scales are often made without any boiler in the accepted sense, the small amount of boiler, smokebox and firebox protruding above the level of the top of the tanks being represented by small segments of suitable tube or flat plate. But this is not really satisfactory in the larger scales, especially when it is remembered that a certain amount of the boiler can be seen below the tanks.

The builder is therefore advised to make up a boiler-smokebox-firebox unit in the usual manner, fit these to the footplate, using a proper saddle, then make the pannier tanks in separate units, as in full-size practice. Each piece, forming one tank, will require two 90 deg. bends, but at a radius that should present no difficulty, bending being done in the vice over a length of metal suitably radiused and a little longer than the tanks. The fitting of each tank unit to the boiler-smokebox-firebox unit certainly needs some care, but it is really only a question of trial and error, a little metal being filed away at a time, until a nice fit is obtained.

Before soldering the pannier tanks to the model, all the details should be attached, such as the tank fillers, the air vents, and the lifting rings. Most panniers also have tee or angle section struts right across the top of the boiler, to hold the tanks in place.

CHAPTER ELEVEN

Tenders

THERE are no great difficulties in making tenders for locomotives of 7 and 10 mm. scales. A minor problem sometimes arises in the fitting of the frames to the body. The practice of permanently soldering the frames to the body may be reasonably satisfactory in the very small gauges, but it is not recommended in gauges "0" and "1".

The body may be made first, starting with the base or floor. It should be noted here that the tenders of some locomotives, particularly the earlier types, were fitted with wheels of rather large diameter, necessitating the cutting of slots in the floor to give the desired clearance. The sides and back made be made separately, or they may be made in one piece: the method to be adopted will depend on the radius of the rear corners. In some modern tenders, the corners are quite square, so that the three piece construction is the obvious answer. If the corners are well radiused, the one piece method may be used, or two pieces could be used, with the join down the middle of the back of the tender.

With the three-piece assembly, it is probably easier to solder the back of the tender to the floor first, using a true block of wood, to ensure that it is upright and parallel to the rear edge of the floor.

Some tender sides have a curved flare to the top edges; some have this flare, plus an additional flat plate or coal plate, typical of most ex-Great Western tenders, and this should always be made as a separate piece, soldered on after the radiused edge has been completed. To produce a nicely radiused top edge to tender sides, it will be found a great advantage if the side is first cut a good deal wider than the finished dimension, so that the extreme edge may be gripped during bending over suitable steel flats and rounds held in the vice. The excess metal is then trimmed off afterwards.

Perhaps the most difficult part of tender body construction occurs in some of the older types of locomotive tenders where the radiused top edge is continued right round the back, as in Fig. 66. To deal with this, the corners of both the sides and the end should first be made somewhat oversize and shaped as in Fig. 67. The sides are then "offered up" to the end, and the surplus metal removed by careful filing with a round or half-round needle

Fig. 65 The flared top edge of the
G.W.R. type of tender can be made in
two pieces (right). The curves are first
bent in a vice and the two pieces soldered
together

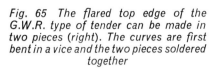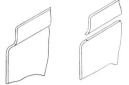

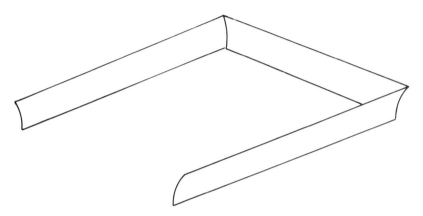

Fig. 66   The top edge of a tender which is radiused all the way
round the sides and back

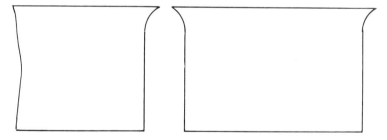

Fig. 67   The corners of a tender's top edge made slightly over-
sized and over-shaped before filing

file. After the sides and end have been permanently fixed to the floor, a
generous fillet of soft solder may be formed over the seam, and this is then
scraped away until a neat and nicely radiused corner has been achieved.

Interior arrangements of tenders differ quite considerably. Some tenders of the earlier types of locomotives did not have the more common and more modern sloping coal "floor", and the rear end of the coal space or recess may be either square or circular. These, however, are quite easy to construct. If the sloping coal floor stretches right across the whole width of the tender, this part is best made in one with the upper "deck" on which is mounted the water filler and generally a water pickup dome and other details. A point to note here is that it is much easier to solder all these details to the underside of the deck (for which purpose they should all be provided with spigots) before fitting the deck/coal plate to the body.

In dealing with the more difficult interiors, as shown in Fig. 68, the deck is made separately, and is added after the coal compartment and tool boxes have been completed and soldered to the body.

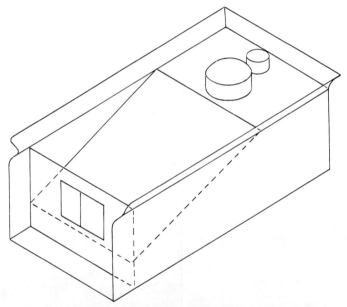

Fig. 68   Interior arrangements, showing the coal compartment and the deck, the latter being soldered on after other details are fitted to it

Tender Frames

There are two or three satisfactory methods of constructing tender frames for gauge "0" and "1" models. The frames proper should be made of brass or nickel-silver about 18 S.W.G. thick for the former gauge and 16 S.W.G. or $\frac{1}{16}$ in. for the latter.

If solid die-cast or brass axlebox and spring castings are to be used, these should be correctly aligned by drilling holes of such a diameter that the bosses on the backs of the castings may just be pushed home, after which, once the axleboxes have been correctly positioned, the bosses can be soldered to the frames. A very light touch of solder at the extreme ends of the spring hangers can then be made on the outside of the frames.

It is often asked how it is possible to soft solder castings made of white metal, whose melting point is only very slightly higher than the solder itself. Apart from strict cleanliness (essential in all soldering operations), the secret is to first flux the joint thoroughly, then to rest the tip of the soldering bit on the frame *just* clear of the part of the die casting to be soldered until the frame itself is hot enough to take the solder. The bit is then quickly moved against the casting and immediately withdrawn again. The bit must be in contact with the casting for only a fraction of a second, but this will be sufficient to form the joint.

To attach the tender frames to the body, short pieces of brass angle may be soldered to the frames, clear of the wheels, and screws put through clearing holes in the angle into tapped holes in the body. It should be emphasised however that the tender floor must be thickened locally to take these screws by soldering brass strip or plate at least $\frac{1}{16}$ in. thick on the top of the floor, and this must clearly be done *before completing the tender body*, while access is still possible.

Another method of construction of tender frames is to use a channel section to carry the tender axles, the outer frames and axlebox/spring castings being merely dummies. The tender wheels, in this method, are therefore running in inside bearings and the tender axles are cut off flush with the fronts of the wheel bosses.

The brass channel used in this method may be purchased ready-made, or it may be built up, in which case $\frac{3}{32}$ in. thick material would be advisable, to give a reasonable length of bearing; alternatively $\frac{1}{16}$ in. material could be used, with the axle holes bushed. In all cases, care must be taken to ensure that all the axles are exactly level, and as in engine frames, it is better to have the centre axle (of a six-wheel tender) very slightly higher than the outer axles than the other way around.

It is often said that the fitting of inside frames to model tenders spoils their appearance in end view. This, however, can almost entirely be overcome by cutting as much of the channel away as possible on each side of the bearings.

Springing of Tenders

The highly-skilled builder will no doubt attempt the fitting of real leaf springs to their tenders, at least in gauge "1", although difficulty will be found in obtaining material of any kind which has sufficient "spring" and at the same time is thin enough for the purpose. Possibly thin steel "shim

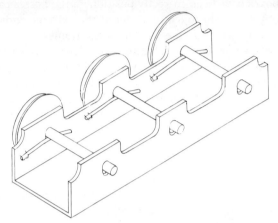

Fig. 69   Method of springing the tender

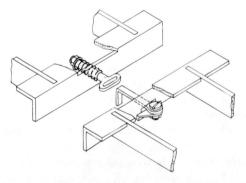

Fig. 70   Spring draw bar and coupling

stock" would be a solution. Another method is to build up the springs from thin sheet metal, but not to rely on them for the actual springing of the vehicle. The buckle is drilled upwards and a very small compression spring inserted in the hole and between the buckle and the axlebox, the latter being recessed just enough to prevent the spring from jumping out.

But a much simpler springing scheme, particularly suitable for gauge "0" tenders, is to adopt the inside channel frame, slotting the axle holes out so as to allow the axles about $\frac{3}{32}$ in. total vertical movement. Suitable spring steel wire is then arranged to bear direct on the axles immediately behind the frames, as in Fig. 69. This wire can be readily clamped by screws and large washers, using some of the screws holding the frames to the body.

The question of the end play to be allowed to tender axles is an important one. We sometimes see tenders in which all the axles have a good $\frac{3}{32}$ in. end play; sometimes even more! This is slovenly and is really quite unnecessary however sharp the radii of the track. A better solution is to allow the outer axles of a six-wheel tender just the minimum end play for easy running, say a bare $\frac{1}{32}$ in. and provide up to $\frac{1}{16}$ in. on the middle axle only, according to the radii likely to be encountered.

With eight-wheel bogie tenders, the bogies are generally built up as for passenger rolling stock, with outside axleboxes and springs. But for free running—especially over track that is none too smooth—equalised bogies are much to be preferred, the bogie frames being pivoted at their centres.

Rigid eight-wheel tenders, as fitted to the ex-L.N.E.R. "Pacifics" and some other locomotives, can be built up exactly as described for six-wheelers, the two middle axles being given sufficient end-play to allow free running on curves.

A spring drawbar between engine and tender is a useful refinement, and assists the locomotive in starting a heavy load. This is easily arranged by providing the coupling hook with a squared shank passing through a square hole in the tender drag beam, the shank being threaded and provided with two nuts and a washer, the spring being placed between the back of the beam and the washer, see Fig. 70.

Tender Details

Many of the earlier locomotive tenders were fitted with coal rails, which were generally of half-round section, though both flat and round sections were used on occasion. The best way of fitting these is first to solder the uprights, of half-round or flat brass wire, to the inside of the tender body. Accurate strips of planed wood of the required thickness are then placed across the top of the tender, as in Fig. 71, and the coal rails, pressed lightly against the wood strips, are then soldered to the uprights. A second or third coal rail may be attached in a similar manner, any excess length of upright being filed flush afterwards.

Other tender details consist of the usual steps, handrails, lamp irons and so forth, and on the top deck are found the water filler (two in some tenders) and if the tender is fitted with water pickup apparatus, there will be the water pickup dome and probably two air vents. There are also generally two lifting lugs at the rear end of the deck.

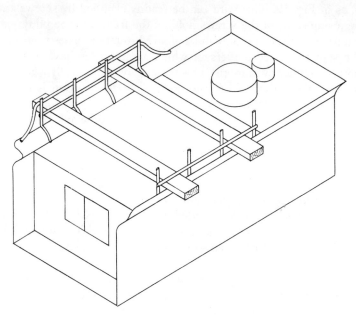

Fig. 71   Method of spacing the coal rails

Although the water pickup dome may be turned from solid brass, this is rather an expensive undertaking as even in 10 mm. scale, the dome may be of quite large diameter. One method of making a pickup dome is to spin it, in the lathe, from annealed copper sheet, about 18 S.W.G. Another method is to cast the dome in white metal or lead, or it may be turned from ordinary mild steel, in which case a threaded fixing spigot should be provided underneath, so that soldering does not have to be resorted to, in fitting it to the tender.

The tender tank filler can be turned from brass or even steel rod, though a built-up construction, using brass or nickel-silver or tinplate sheet, is to be preferred, as this makes the soldering of details such as hinges and handles much easier. For fitting the filler to the tank top, a "floor" may be added of rather thicker metal, this being tapped 6 or 8 BA, for a securing screw put in from underneath.

The air vents can be straightforward brass turnings, provided with a thread at the bottom end, for easy attachment to the tender, while tool-boxes, which are almost invariably found on locomotive tenders, can be built up from thin sheet metal.

A final touch of realism for the model tender is the filling of the coal space with some actual coal. Perhaps the best way of doing this is to first shape up a piece of softwood to fit the coal space fairly closely, this is then painted flat black and covered with a spinkling of coal broken up finely according to the scale of the model. Any good tube adhesive should be suitable to hold the coal down.

CHAPTER TWELVE

Detail Work: Riveting

Handrails

MANY MODEL locomotives are spoilt by oversize or incorrect handrail knobs. Although brass split pins make passable knobs, if fitted with a small washer, for gauge "0" models, when we come to gauge "1" a proper turned knob is desirable. A turned handrail knob can be threaded 10 or 12 BA and nutted on the back wherever the nut is accessible.

On tapered boilers, handrail knobs are often of different lengths, and on some locomotives, the handrail alongside one side of the boiler is not fitted direct to the boiler, but is attached to the blower pipe. Some engines had the blower pipe support and handrail knob combined. In this case, the blower pipe is first soldered to its supports and then drilled on the outside of each support with a fine drill, and smaller handrail knobs inserted in the holes. For the handrail itself, nickel-silver or stainless steel wire may be used, bearing in mind that the full-size handrails varied in diameter from 1 in. to $1\frac{1}{2}$ in.

Check or Clack Valves

On some locomotives, the check or clack valves for water feed to the boiler were arranged on the sides of the boiler barrel and were therefore quite noticeable. A typical check valve is shown in Fig. 72, from which it will be seen that there is a flanged connection for the feed pipe and another at right-angles for attachment to the boiler barrel.

On the later Great Western engines, and the Stanier engines of the L.M.S., top feed was used. On the G.W. locomotives, the check valves were arranged on each side of the twin safety valves, the whole being covered by a conical cover, so typical of this railway. Apart from some of the earlier examples, the Stanier top feeds were quite separate, fitted to the top of the boiler barrel in front of the steam dome, the safety valves being of the Ross "pop" type, fitted near the cab. The top feed for gauge "0" and "1" models is best filed up from the solid brass, though the underside may be fly-cut to match the diameter of the boiler barrel at the appropriate point, as described in Chapter Nine.

The making of the Great Western combined safety valve cover and top feed always presents a difficult problem to the builder. Perhaps the best way to tackle it is to turn up the safety valve cover first, saddling it to match

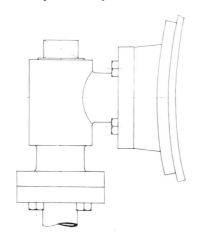

Fig. 72   The check valve

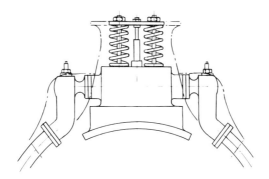

Fig. 73   The combined safety valve
and top feed typical of the Great Western
Railway locomotives

the boiler barrel, then cut a cross-wise slot through the underside, deep enough to take the top feed. This is then filed up from a separate piece of solid brass, fitted to the cover and soldered to it. The underside of the complete fitting may then need a little final careful filing, and it is then drilled and tapped to take a fixing screw, which can be put through from the inside of the boiler.

Lamp Brackets

It is really worth while fitting lamp brackets to gauge "0" and "1" loco-motives. To prevent them getting bent during handling, they should be made slightly thicker than scale, and preferably of nickel-silver. Do not solder them direct to the running board or smokebox door, but drill a hole, push the bracket home, and, where possible, solder it on the back.

Caledonian Railway locomotives were fitted with additional lamp irons on the sides of the cabs, while L.N.W.R. lamp brackets were quite different, having a square recess to take a spigot underneath the lamp.

Steps

Like lamp irons, the steps on model locomotives are easily bent or broken off. To avoid this trouble, they should be made of nickel-silver, at least 20 S.W.G. for gauge "0" or 18 S.W.G. for gauge "1". In addition, after soldering the step to the running board, build up a little fillet of solder on the back and also solder a piece of brass or nickel wire at an angle to form a strut as in Fig. 74.

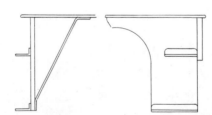

Fig. 74   Strut providing support for the steps

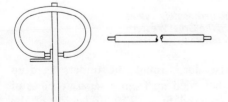

Fig. 75   Silver steel rod for temporarily clamping the step to the backplate while soldering

As it is generally easier to solder the middle or upper step to the backplate before fitting the assembly to the model, this step usually becomes unsoldered while soldering the backplate. To avoid this trouble, bend up a length of $\frac{1}{8}$ in. silver steel rod so as to act as a temporary clamp while carrying out the soldering operations, as in Fig. 75.

Vacuum brake pipes

Vacuum brake pipes are prominent features on passenger locomotives, especially at the front end. Some engines were Westinghouse fitted and some were dual fitted (both vacuum and Westinghouse). The Westinghouse pipes were always considerably smaller than the vacuum.

We seldom see really convincing vacuum brake pipes, the main difficulty being to obtain something suitable for the hose.

It may be difficult to find rubber tube small enough in diameter for gauge "0" models, or even for gauge "1". However, one possibility lies in certain kinds of flexible electric wire, which has a rubber covering, the braided copper wire being first withdrawn.

The "standard" of the brake pipe requires careful bending; if brass wire is used, it should first be turned down to form the part which is to be covered by the rubber tube, as the outside diameter of the hose is generally only slightly greater than the standard. It is then heated to red and the sharp bend at the top made while holding it in two pairs of pliers, one "flat" and one "round-nose".

Most passenger locomotives were fitted with steam heating, so a steam heat pipe should be attached to the rear of the tender, or in the case of tank engines, at both ends. Some passenger tender engines (notably on the old Great Eastern Railway) were fitted with steam heat connections at the front of the engine in addition to the rear of the tender.

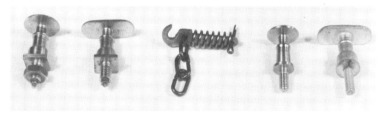

Plate 33   Gauge "0" buffers and couplings, by Messrs. Bond's

Buffers

Buffers are another fitting which are seldom reproduced accurately in the

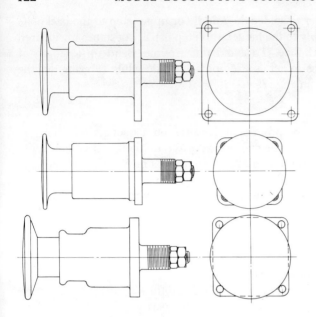

Fig. 76　Styles of locomotive buffers

Fig. 77　To prevent oval-head buffers from turning: top, a pin may pressed into a slot underneath; centre and below, a light metal strip may be attached to the rear of the buffer spindles

Fig. 78　A screw coupling

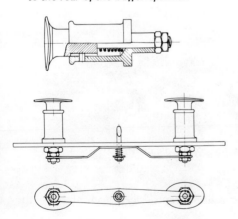

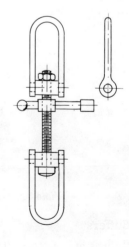

*Plate 34A Locomotive
fittings: Buffers
Right, above: Ross
"pop" safety valve
Centre, left: Ross
"pop" safety valve
and a tender
water filler*

Smokebox doors

*Plate 34B Single-link,
three-link and screw
type couplings, and
a Southern Railway
driving wheel*

smaller scales, the most common fault being the lack of radius at the back of the buffer head, between the head and the shank. Fig. 76 shows several designs of locomotive buffers.

Spring buffers are certainly desirable in 7 and 10 mm. scales; the springs being light steel compression springs inside the buffer stock, or in the case of some of the Gresley (L.N.E.R.) designs, behind the buffer beams.

Oval-head buffers sometimes present a problem, owing to the necessity to prevent the heads from turning. Two methods may be tried. Firstly, the buffer stock may have a small longitudinal slot milled underneath, and a pin pressed into the shank, the pin being made a good fit in the slot. Secondly, the rear ends of the buffer spindles may be attached by a light flexible wire to some convenient point on the engine so as to permit the spindle to move in and out, but to prevent the spindle from revolving. The two methods are shown in Fig. 77.

Plate 34C   Further locomotive fittings by Messrs. Bond's. Top left, engine step; right, a buffer; below, three types of tender axlebox and spring casting

Couplings

All passenger engines and mixed-traffic engines and some freight locomotives were fitted with screw couplings. These are certainly fiddling things to make, especially in 7 mm. scale. The shank of the hook may be made rectangular, passing through a similarly-shaped slot in the buffer beam. A compression spring, washer and split pin should be fitted to the shank behind the beam, the spring assisting the starting of the train.

The shape of the hook should be watched carefully; it is seldom made correctly even in the larger scales. Fig. 78 shows the correct shape of one full-size design and also gives details of the links and shackles.

Beading

A model locomotive looks very bare without correct beading. Most locomotives had beading around cab side-plates and windows and around wheel splashers, but it is not an easy matter to fit this neatly.

Perhaps the best way to go about it is to first make up two miniature clamps, using $\frac{1}{8}$in. square silver steel, the screws being 8 or 10 BA. These should be allowed to go rusty (easy enough in a damp atmosphere!) so that the solder will not adhere to them.

The beading, which should be half-round brass wire, must be quite straight before commencing operations, and a length of at least one inch more than is required is cut off.

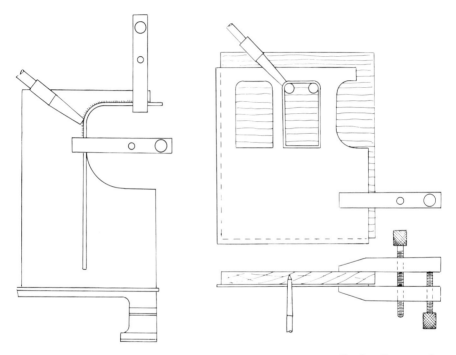

Fig. 79 Soldering the beading with aid   Fig. 80 Soldering the beading round a
        of two clamps   cab window

Tin the surface to which the beading is to be applied, using the minimum of solder so that the beading will fit closely to the plate. Start by attaching the beading to one end by fixing the two miniature clamps at a distance apart such that the tip of the soldering iron can be inserted comfortably between them (See Fig. 79). If a tiny drop of liquid flux is now applied, followed by the soldering iron, a sound joint will be produced.

The outer clamp is now moved close to the second clamp, that is, covering that part of the beading which has been soldered. The part of the beading exposed is next soldered.

One clamp only can now be used, applying it to the part which has just been soldered, the beading being bent by hand in stages.

In certain positions, it may not be possible to use even these very small clamps; in such cases, it is possible to hold the beading down in the required position by a piece of stripwood to start the operation, the wire being first "tacked" at intermediate points, the gaps being soldered afterwards.

Cab window beading generally involves some very sharp radii in the corners. Start by soldering the beading along one straight edge, then bend it sharply around a scriber or something similar, which is pressed hard up against the corner of the window as in Fig. 80.

Where the cab lookout windows or "spectacles" are circular, the beading may first be wrapped several times around a length of rod of somewhat smaller diameter than the cab window, after which the rings can be snipped off with wire cutters. Those with lathes will no doubt turn up suitable rings from brass tube of the appropriate diameter and thickness.

Cab Fittings

Cab fittings add the final finishing touch to the "super-detail" electrically-powered model locomotive.

The fittings in the cabs of the early steam locomotives were comparatively simple, the main items being the regulator, the blower valve, a water gauge, a pressure gauge, the reversing lever (or screw) some form of brake and the fire door.

Later such things as steam and vacuum brake valves, vacuum gauge, steam heat valve, whistle valve, cylinder drain cock controls, firebox damper controls, etc., etc., were added, so that the cab layout became most complicated.

Even in a 10 mm. scale model, it would be very difficult to reproduce reasonably accurately all the fittings in the cab of a modern locomotive. It is therefore far better to try to model properly the more important fittings, such as the regulator, the firedoor, the water gauges, the pressure gauges, the driver's and fireman's seats and of course the reversing gear.

The cab fittings are best built on a separate plate, bent at right angles, the upright part representing the backhead (back of the firebox) and the horizontal part of the footplate, to which can be attached a hinged "fall plate". A beading may be fitted around the corners of the backhead and some representation of the wash-out plugs and "hand holes" can be fitted. If

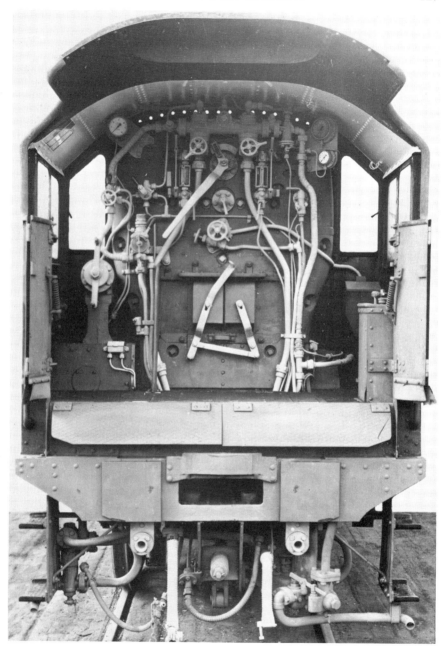

Plate 36   Cab fittings: B.R. "Britannia" Class

Plate 35   (Overleaf) Cab fittings: Class 5 M.T. No. 5268

Plate 37   Cab fittings: S.E.C.R. 4–4–0 Class D

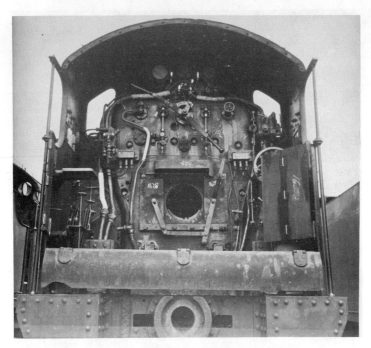

Plate 38   Cab fittings: S.R. 2–6–0 Class U
Plate 39   Cab fittings: Midland 4–4–0 Compound

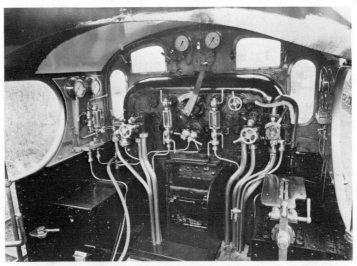

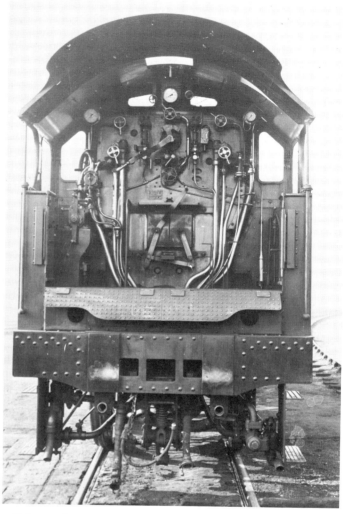

Plate 40 Cab fittings: "Royal Scot", L.M.S.

Plate 41   Cab fittings: L.N.W.R. 0–8–0 goods

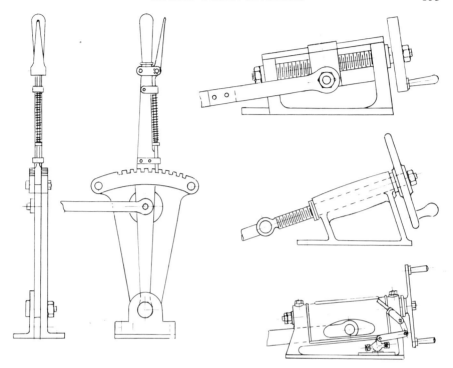

Figs 81A, B, C and D Various types of cab reversing gear. Left, Lever type. Top, right, typical locomotive screw reverser. Centre, right, an early L.N.W.R. screw type. Below, right, a screw reverser with extra lever for operation by steam or compressed air

possible a photograph of the prototype cab should be studied, and to assist builders, photographs of the cabs of some well-known locomotives, some modern, some not so modern, are included in this book.

Making the circular dial of the pressure gauge is sometimes a problem. One method is to draw it out, complete with needle, in Indian ink on stiff white paper or card, when it can be photographed and reduced to the desired size for the model.

Copper wire of various gauges can be used to represent the many pipes seen in the locomotive cab, but do not forget the various flanged joints, for which very small washers can be used.

Various types of cab reversing gear are shown in Fig. 81. Note the distinctive Gresley type (Fig. 81E) with vertical screw, as fitted to his three-cylinder locomotives—not the easiest to turn, as the author once found out to his cost.

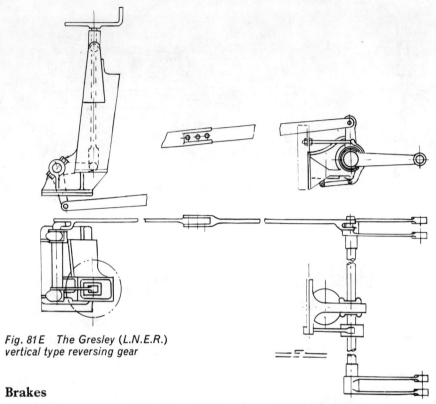

Fig. 81E The Gresley (L.N.E.R.)
vertical type reversing gear

Brakes

Model locomotives to 7 mm. and 10 mm. scales look rather bare without brake gear, especially those engines with high running plates where most or all of the driving and coupled wheels are exposed.

With electrically driven models, it is a good plan to make the whole brake rigging readily detachable from the frames of the locomotive. This can be done by the "spring on" method, well known to "00" gauge enthusiasts; the idea can be readily understood from Fig. 82.

In the early days of the steam locomotive, the brake blocks were made of hardwood, but this was soon superseded by cast iron. The brake blocks are generally slotted down the back and attached to steel hangers, which are supported by either a shouldered pin or a small casting, bolted or riveted to the frames.

The brake rigging, that is the pull rods and cross-connecting beams, may be either compensated or uncompensated, and it may be either inside or outside the coupled wheels.

In full size practice, the criticism was sometimes made of the normal brake arrangement that the pressure of the brake-block on the wheel tended to force the driving or coupled axle to one side, and if there was any appre-

Fig. 82
Showing
how model
locomotive
brake gear
can be
"sprung"
into
position

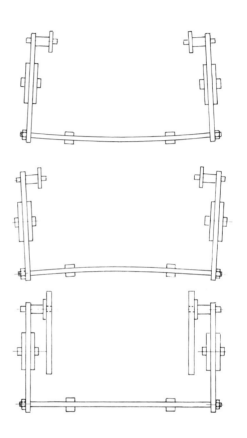

ciable wear in the axleboxes and horn faces, considerable stresses were thrown on the coupling rods.

This disadvantage was overcome in the ex-Southern Railway Bulleid "Pacifics" by an ingenious system of "clasp" brakes in which each wheel was provided with two brake-blocks, one on either side, each pair of adjacent blocks being pivoted together with one hanger, as in Fig. 83. A somewhat similar arrangement was used by Stroudley on the L.B.S.C.R., and by Johnson, Deeley and Fowler on the Midland Railway.

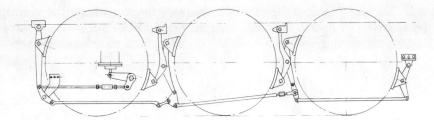

Fig. 83   The system of "clasp" brakes used on the Bulleid "Pacifics"

At one time, the bogies of some Great Western locomotives were also fitted with brakes, but as these proved very difficult to keep in proper adjustment, the idea was abandoned by Mr. Collett. In any case, bogie brakes would hardly be worthwhile on a model apart from a "glass-case" exhibition model, as they are very inconspicuous.

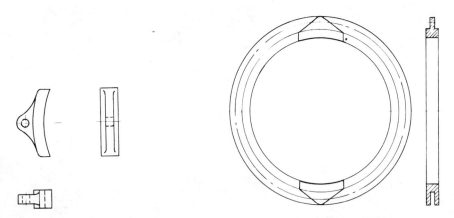

Fig. 84   Three views of a brake block for a double hanger

Fig. 85   The "ring" method of making brake blocks with a lathe

Express locomotives were generally fitted with vacuum brakes (or Westinghouse, or dual brakes) and goods engines with the steam brake, though there were exceptions to this. Tank engines were fitted with hand brakes, generally a vertical screw-type, so that if coupled to a steam brake, a slotted link would be fitted so that the two systems could operate independently of one another.

The making of the brake blocks for the model locomotive can be tackled in two distinct ways: from a solid flat strip of brass of the required thickness, or by turning a ring in the lathe and cutting lengths from this. In the first method, mark out as many brake blocks as are required, drill the holes, cut off and file to contour, using a half-round file to finish the working surface (that next to the wheel). If the brake block is for a single hanger, the back will have to be slotted.

This can be done by making a saw cut, choosing a saw of a thickness to match as nearly as possible the thickness of the hanger.

For those with lathes, a thin slitting saw can be set up on a stub mandrel or between centres, but in this case, it is better to leave the strip intact, slot one end, using the remainder of the strip as a "holder" (it can be clamped under the lathe toolholder) cut off the first brake block, slot the end again, and so on, until all the blocks have been done, when they are filed to shape.

If the brake block is for a double hanger, file away both sides of the rear part as shown in Fig. 84.

The "ring" method of making brake blocks is probably the better one for lathe owners. The inside of the ring can be bored out to the exact diameter of the wheel treads, while the outside can either be slotted with a narrow parting-type of lathe tool, if for single hangers, or reduced on either side, as in Fig. 85, if for double hangers.

Westinghouse Brake Air Pump

The air pump of locomotives fitted with the Westinghouse brake is often

Fig. 86 The Westinghouse brake air pump, showing the lubricator (top), reversing valve and air cylinder (below)

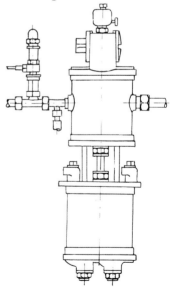

a very prominent feature, yet it is seldom modelled correctly. Fig. 86 shows the external appearance of this pump. The upper cylinder is the steam cylinder, which has a reversing valve and a small lubricator on the top, while the lower cylinder is the air cylinder, the piston rod being common to both cylinders.

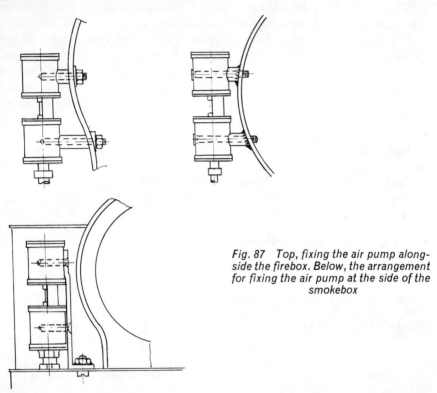

Fig. 87   Top, fixing the air pump along-side the firebox. Below, the arrangement for fixing the air pump at the side of the smokebox

The action of the air pump is controlled by a governor, which admits steam to the pump when the air pressure falls to a certain predetermined amount, and shuts it off automatically when the pressure has risen to the working figure.

In the model, the main body can be a straightforward turning, comprising the two cylinders, the intermediate section being filed back and a rod inserted to represent the piston and glands. The governor can be represented by a separate turning, soldered to the upper cylinder, and the various pipes by suitable copper wires.

Fixing the air pump to the model locomotive will depend on its position.

On some lines, the pump was situated alongside the firebox, in which case two small screws could probably be put through from the inside of the firebox, to protrude right through the two cylinders; the screws are then soldered and filed flush.

If the pump is arranged at the side of the smokebox, or in the case of tank engines, just in front of the side tanks, perhaps a better method is to first screw a narrow strip of metal to the back of the pump, using countersunk screws into tapped holes in the pump cylinders. The bottom end of this strip is then bent at right-angles, and a nut soldered to this. A small screw can then be put into this from the underside. The two methods of mounting are shown in Fig. 87.

Riveting

Many full-size locomotive engineers, especially on the Great Western Railway, made great use of snaphead rivets in the construction of their engines. For instance, there were rows of snaphead rivets around the front and rear of the smokebox, rows along the running plate and its stiffening angle, on the cab sides and front and on the buffer beams. The tenders, too, showed a great number of these rivets.

The super-detail or exhibition model, therefore, would not be complete without a representation of this rivet detail.

The great thing in adding snaphead rivets to a model is to keep the "heads" to the scale size; nothing looks worse than over-scale rivets, especially if they are arranged at the scale spacing, when their size shows up even more. The author well remembers an attempt at building a detailed G.W.R. Pannier tank locomotive while in H.M. Forces just after the War, when no proper workshop facilities were available. Having no jig, or means for making one, for producing the larger number of rivet heads, it was decided to drill holes whenever a rivet was required, and to insert in each a very fine brass pin (actually a "00" gauge track pin) from the back, the head being soldered. The front of the pin was then snipped off and after all pins had been treated similarly, they were well rubbed down with emery cloth, rubbing equally in all directions, so that the ends of the pins were nicely rounded. The final effect was encouraging, but in view of the enormous amount of work involved, the method can hardly be recommended!

The best way to produce rivets is to make a jig. This should be made from a bar of steel, preferably tool steel. A row of rivets is carefully marked out, drilled and finished with the rounded end of a fine punch. The exact depth and diameter of the depressions in the jig must be found by experiment beforehand, using scrap pieces of the same metal as used for the superstructure of the model.

A removable guide strip will be required, so that the distance between the line of rivets and the edge of the plate can be altered as required. It is not essential to have the jig as long as the plate to be riveted, but it should have at least a dozen depressions so as to give reasonable stability.

The punch must be made to suit the "punch-bush", which is attached rigidly to the guide strip. It will be noted that the punch must have a parallel part (A in Fig. 88) which must be made a good fit in the bush B.

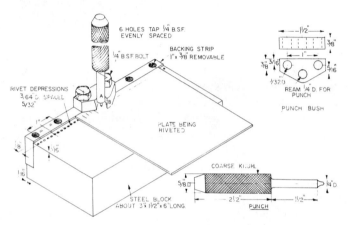

Fig. 88   Using a jig to produce dummy riveting

To use the jig, the plate only has to be marked out for the very first rivet required. The plate is aligned under the bush and the first rivet formed by a light tap with a hammer. The plate is then slid along, keeping its edge hard against the guide strip, until the rivet which has just been formed drops into the next depression, when the next rivet is formed, and so on. Thus the spacing of the rivets will exactly match the spacing of the depressions in the jig.

For the best results, the jig should be hardened and polished before use, as should the punch.

CHAPTER THIRTEEN

Outside-frame Locomotives and Special Types

SOME of the most attractive full-size steam locomotives were fitted with what are rather loosely called outside frames, though in actual fact, most of these engines had double frames, that is frames both inside and outside the driving and coupled wheels.

In model form, a very good representation can be produced by making the outside frames dummies, which may be permanently attached to the super-structure. The only real snag to this scheme is that, if the inside frames with the motor are to be readily detachable, the horn slots of the outside frames must be left open, without any hornstays. In gauge "1" work, it is possible to fit hornstays which are held to the outside frames by very small screws and nuts, say 12 BA. In this case, axleboxes may also be provided for the outside frames, but as it is rather difficult to line these up exactly with the inside axleboxes, at least in the smaller scales, the outside axleboxes can be made quite an easy fit on the extensions of the axles.

Another method of fitting the outside frames is to use spacing pieces between inside and outside frames and fix the whole together by rods threaded

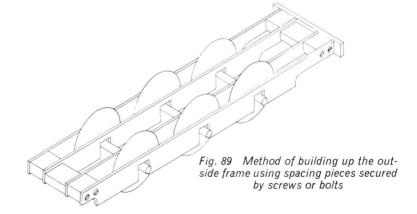

Fig. 89   Method of building up the outside frame using spacing pieces secured
by screws or bolts

at each end, with nuts on the outside of the outside frames. This is shown in Fig. 89.

The driving and coupled axles for outside frame locomotives need special treatment. Probably the best way is to use axles with the wheels a press fit, and to shoulder the axles down where they pass right through the wheels. The axles are then shouldered down once again to take the outside cranks, as in Fig. 90.

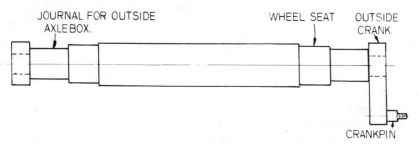

Fig. 90   Shouldering down an axle for outside framed locomotives

The outside cranks should be made of mild steel and made a press fit on the ends of the axles, but if there is any suspicion of slackness here, they could be soft soldered as well. The cranks must be drilled in a simple jig to ensure that they are all of exactly the same throw.

"Quartering" of the outside cranks, to ensure that they are all at the same angle relative to those on the other side of the locomotive, (they should be at 90 deg. to one another, with the right-hand cranks leading) is best done with the coupling rods. The rods are first made a rather close fit on their crankpins, but after quartering, the holes should be opened out sufficiently to allow free running.

The crankpins should be made of silver-steel, a press fit in the outside cranks, threaded on the end, and fitted with nuts and washers.

American Locomotives

When modelling typical British locomotives, we generally encounter a running plate which, for modelling purposes, is usually considered as a part of the body or superstructure, external parts and details such as outside cylinders, valve gear, brake gear, sanding gear, etc., being attached to the chassis.

In most American locomotives, there is no easy line of division between the superstructure and the chassis. The running boards are usually arranged quite high up and are attached to the boiler and have no connection with the pilot beam (the equivalent of the British buffer beam). The frames too are quite different, being of the bar type, instead of the British plate type.

When modelling the large American locomotives, it is generally most convenient to arrange the motor in the firebox, as this is always much larger than in British practice. The advantage of doing this is that there is no need to cut into the boiler barrel at all; the barrel can be completely circular,

Plate 42   *View of an American model locomotive. Note the pilot or "cowcatcher"*

except possibly the extreme rear end of the barrel just in front of the throat-plate of the firebox, where the driving shaft of the motor may protrude.

In most American locomotives, the saddle and cylinders were cast as one piece or occasionally in two parts joined on the centre-line. In the model, it is probably easiest to follow suit, making the saddle a permanent part of the chassis, and fixing the smokebox down on it by a fairly substantial screw put through the saddle into the bottom of the smokebox, which should be thickened locally so that it can be tapped.

The pilot beam will be attached to the bar frames, and this can be done by screws put through the front ends of the frames in the usual way.

At the rear end of American locomotives, the ashpan is quite a prominent feature, as this spreads out to approximately the full width of the engine. The ashpan therefore might well be attached to the chassis, and the edges of the firebox may then locate inside its edges, as in Fig. 91.

The rear of the superstructure can generally be held down to the chassis by a screw put through a stretcher close to the rear end of the frames, into the cab floor, on top of which a nut can be soldered, this being covered by a false floor, forming the backhead of the boiler and carrying the cab fittings, as described in the previous chapter.

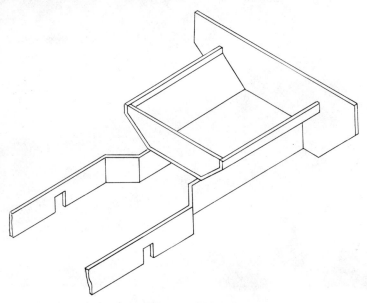

Fig. 91   Diagram of an ashpan for an American locomotive

A prominent feature on American locomotives is the pilot ("cowcatcher") which is rather a difficult item to make. A good plan is to make up a simple jig, using a wooden base. The lower "frame" is cut from brass strip or square bar to the shape required, but left considerably longer at each end than is actually required, so that the ends can be held down on the wooden base by drawing pins or panel pins. Notches are filed along the top of this frame, and a similar row of notches is made along the bottom frame.

Lengths of suitable round or square wire are now laid across in the notches, the joints soldered, and the ends trimmed to shape. Finally, another strip, of flat section, is soldered across the bottom frame, The sequence is shown in Fig. 92.

Generally speaking, the pilots of early American engines were longer and more pointed, while the most modern engines were fitted with pilots built up from bar metal, or made from castings.

Plate 43   Typical American "cowcatcher" from casting

The boiler fittings of American locomotives had some similarity with their British counterparts, though they generally had one or two sandboxes mounted on the top of the boiler, which closely resembled steam domes externally. Stacks (chimneys) were generally of the "stove-pipe" type, except in the early wood-burners, where the stacks were of enormous proportions, built up from sheet metal.

Two other items always found on American locomotives were the bell and the headlamp. The bell itself will have to be turned in the lathe, but the

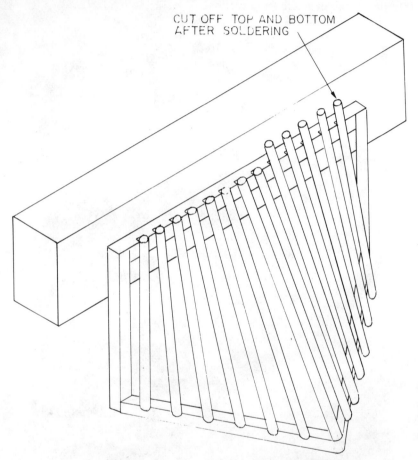

Fig. 92   A "cowcatcher" made with the aid of a wooden jig

support could either be bent up from flat wire, its edging having been previously rounded off by filing, or it could be cut from the solid. A spigot should be provided at the bottom for securing to the smokebox or boiler.

The oil-burning headlamps of the earlier American locomotives were often huge affairs and generally highly ornamental. The lens should be represented by a glass bead or by a disc of Perspex. The body can be built up from thin sheet metal, and the bracket fretted out, bent to shape and soldered to the "floor".

Plate 44   Another type of pilot or "cowcatcher" seen on a 3½-inch gauge American 4-6-0, by A. W. Leggett

Power Tenders

When modelling some of the small-boilered earlier locomotives, it is sometimes found that it is not possible to find room for the motor in the locomotive itself, without a large part of the motor being exposed to view. This sometimes happens with engines with high-pitched boilers such as the more modern 0-6-0's and 0-8-0's.

There are two solutions to this problem: put the motor in the tender but fit it with a flexible shaft, or a shaft with universal joints, driving the locomotive wheels through a longitudinal shaft and worm gears; or have the motor in the tender but driving directly on to the tender wheels.

The problem with the flexible shaft between engine and tender is to allow sufficient flexibility to enable the model to negotiate curved track. Whenever possible, the driving shaft should be arranged on the longitudinal centre-line of the model, as it is here that the sideways movement between engine and tender is at its minimum.

The motor in the tender is usually arranged on its side, and in many cases, it is found that the height of the armature spindle above track level, even if the motor is placed as low as possible without fouling the tender wheels, is too high to allow a direct drive to the worm shaft in the locomotive. If this should prove to be the case, an intermediate pair of flat gears can be introduced.

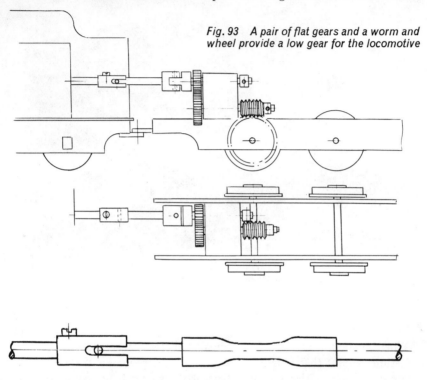

Fig. 93   A pair of flat gears and a worm and wheel provide a low gear for the locomotive

Fig. 94   Flexible drive using rubber tube and metal coupling

These additional gears may serve another function, as advantage may be taken of them to alter the gear ratio. As power tenders are generally required for slow running goods and freight locomotives requiring a low gear, the upper pinion of these intermediate gears can be made smaller than the lower.

Fig. 93 shows the arrangement, the drive including a simple universal joint and a telescopic joint, the pair of which should take care of longitudinal and lateral movement between engine and tender.

Another method of arranging a flexible drive between engine and tender is to use a length of rubber tube or rod, as in Fig. 94, but in gauges larger than "00", some form of metal coupling should be fitted at each end of the rubber, to ensure a positive drive. The free length of the rubber should not be too great, or it will merely twist up instead of transmitting the power through to the engine.

An important point to note with flexible shaft drives is that the motor must be of such a design that the armature shaft is brought right through the outer bearing, so as to take the telescopic or universal joint. A motor in which the ends of the armature spindle are enclosed between bearings of the "pin-point" type or bearings of the ball-thrust type would not be suitable without modification. However, such motors can generally be modified to allow the armature spindle to protrude through a parallel bearing, or an extension shaft coupled to the spindle can be fitted.

Where this modification is carried out, it is most important to fit a collar on the inside of the outer bearing to limit the end-play of the armature spindle, particularly where the bearing at the magnet end is of the end-thrust type, when the end-play must be reduced to the absolute minimum without any tightness at running temperature. It will be appreciated that parallel bearings will be quite satisfactory in a tender motor as the worm and worm-wheel have now been moved to the locomotive chassis, where the end-thrust of the worm gear must be taken care of.

Where the outer motor bearing has to be altered from an "end-thrust" type to a parallel type, great care must be taken to ensure that the alignment of the armature in relation to the magnet pole pieces is not upset.

As to current supply, if the locomotive is well weighted, the pickups can be fitted to the locomotive in the usual way, an insulated flexible wire being taken through to the tender.

As previously mentioned, the alternative to the flexible drive between tender and locomotive is to couple the motor in the tender to its own wheels. The current collectors can then be fitted to the tender, which becomes a completely self-contained power unit.

It is essential in a "power-tender" that the drive be transmitted to all the wheels, otherwise bad wheel-spin is sure to occur. As outside coupling rods cannot normally be used, chain and sprockets over all axles can be considered. But the best way is probably to use a longitudinal shaft carrying worms above each axle, to which are fitted corresponding worm-wheels. The drive from the motor can then be arranged by a pair of spur gears. Fig. 95 shows this scheme.

The longitudinal shaft will require "end-thrust" bearings at each end, and the motor can be laid on its side, as before. To ensure plenty of adhesion weight, strips of lead can be placed over the top of the motor, and this can be

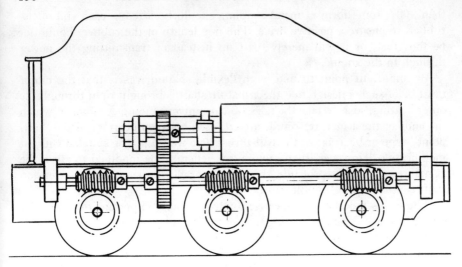

Fig. 95   The motor coupled to the wheels of the tender, making
a completely self-contained power unit

covered by the usual top deck with a thin layer of coal in the coal space.

The longitudinal shaft arrangement prohibits the springing of the tender wheels unless a rather elaborate system of floating worm gearboxes with universal joints between each is adopted. The chain drive scheme would not however prevent the wheels from being sprung.

Power-tenders are best built with inside frames, the outside frames with dummy axleboxes being fitted in the usual way to the tender floor.

CHAPTER FOURTEEN

Painting and Lining

IT is extraordinary how many otherwise fine models are spoilt by indifferent painting and faulty and inaccurate lining and lettering. It is often said that "paint covers a multitude of sins"; while this might be true of large size work, it is certainly not true of small scale models of any kind. Mistakes, blemishes and the like nearly always show up through the paint.

The author is of the opinion that the reason why so many models are poorly finished is undue haste. Good painting takes time, both in the actual painting process and during the equally important preparation.

The first thing to do is to clean as much as possible of the model with the finest grade of emery cloth, working with a circular motion rather than up and down. This should be followed by careful rubbing with jeweller's rouge, pumice powder or even "Vim".

After this, the model should be washed in hot water containing a little washing soda, at the same time carefully brushing it all over with an old toothbrush or something similar. It is then thoroughly rinsed in clean hot water and allowed to dry, after which it should not be touched with the hands more than is absolutely necessary.

Paints

It is now time to consider the type of paint to be used. Paints suitable for model work may be broadly divided into two main classes: (1) oil-bound pigments, which are slow drying and can either be sprayed or applied by brush; (2) cellulose and synthetic paints, which although they can be applied by brush, generally give the best results when sprayed.

It is sometimes claimed that the best results on model locomotives can only be obtained by spraying. There is no doubt that much more skill and practice is required before good results can be obtained from brush painting rather than spraying.

If it is decided to adopt spraying, it should be realised that if the complete equipment has to be purchased, this is likely to be a costly business. However, for those with suitable workshop equipment, the making of a proper high-pressure spray gun is not difficult. (Drawings of a suitable spray gun can be obtained from the publishers of this book.)

In most cases, the amateur will purchase his supply of paint for model locomotive work from the model supply houses, as in this way the correct

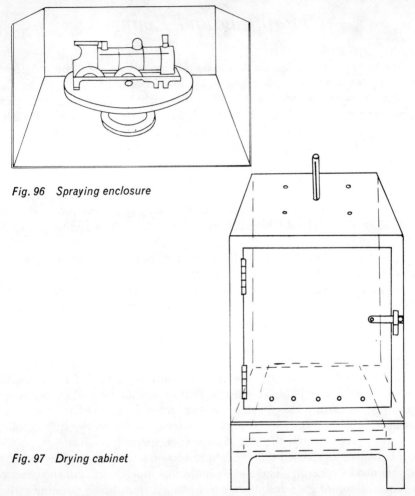

Fig. 96   Spraying enclosure

Fig. 97   Drying cabinet

colours will be obtained for the railway being modelled. If cellulose paint is required, this will mean careful matching against the paints, though in the case of models of the "pre-group" railways, the paint will have to be mixed from first principles. Care should therefore be taken only to mix paints of the same type and from the same manufacturers. This precaution will also apply to the necessary thinners.

It is sometimes asked whether the various "aerosol" paint sprays now on the market can be used for model locomotive painting, as great varieties of colours can be obtained. Unfortunately, however, these paint sprays cannot be relied upon for such fine work, their construction being of insufficient precision for reliable work. The "spray" of paint from these aerosols is generally too heavy and too coarse for model work, and the delivery pressure too unreliable.

For spraying model locomotives, or in fact any kind of model of comparable size and construction, a three-sided enclosure should be made up from sheet metal and a small turntable should be made or acquired, on which the model can be mounted—an old gramophone turntable would be ideal. The set-up is shown in Fig. 96.

Every precaution should be taken to avoid dust settling on the model; this is not quite so important in spraying as the paint dries very quickly, but in hand painting, the only satisfactory way of dealing with the dust problem is to make up a drying cabinet, as in Fig. 97, into which the model can be placed immediately after painting. Such drying cabinets are always adopted when stoving enamels are used, but even with ordinary paints a very slight heating to aid drying of the paint is an advantage.

Hand-painting

Generally speaking, a model locomotive should be given three thin coats of paint, followed by a thin coat of varnish after the lettering and lining, if any, has been completed. The first coat should be an undercoat, of a lighter colour than the finishing coats.

The paint should be applied with sufficient on the brush to ensure proper flow over the work, but the brush should not be dipped too deeply into the paint. It should be decided beforehand what is thought to be the best order to paint the various parts of the model and no attempt should be made to cover the whole locomotive in one go. In fact, it is advisable to paint only those surfaces which cannot be laid horizontally first, and then tackle each horizontal surface in turn, allowing each area to thoroughly dry before proceeding to the next. Then if any paint has crept over on to an area which is to be painted at a later stage, it can be removed by careful application of fine emery cloth.

For instance, the boiler of the model might be tackled first, brushing this around the barrel, and not longitudinally. When this is thoroughly dry, the model could be set up on end, buffer beam uppermost, and the front of the cab painted. To do this, the locomotive should be firmly supported, and a wooden block or something similar should be provided of a height corresponding to the model, on which to rest the arm while painting.

The next item to receive attention might be the cab sides, or in the case of tank locomotives, both the cab and the tank sides, in turn. Black parts, such as the footplate, tops of splashers, smokebox and chimney and cab roof, would receive attention at a later stage.

Painting should always be carried out in a room which is reasonable well ventilated, but without draughts, and the contents of the room should be disturbed as little as possible beforehand, to prevent dust rising. The room should be at a temperature of about 70 deg. F, and reasonably dry weather should be chosen whenever possible.

In hand painting, it is very difficult to avoid the paint gathering in drops or tears around the details; the brush should however be drawn away from details and also away from edges, when giving the final strokes.

Fig. 98   A piece of wire twisted round the brush and resting on the top of the glass or jar keeps the hairs and the lower half of the metal holder suspended in spirit when the brush is not being used

When applying the first coat on metal, it is generally found that the metal shows through in places; but no attempt should be made to completely cover such places with the first coat, as the second and final coats will deal with them adequately.

Brushes

For painting model locomotives of 7 and 10 mm. scales, a flat sable brush about $\frac{3}{8}$ in. wide will be suitable. A good quality oxhair brush makes a reasonable substitute for sable and will be considerably cheaper. A selection of small round sable brushes should also be acquired, a "00", "0" and "1" size being very useful for the fine details, lettering, etc.

Good paint brushes are very expensive, so great care should be taken of them. If used on water or poster colours, they should be washed in luke-warm water immediately after use, shaken thoroughly and the hairs then drawn straight before being put away.

Brushes that are being used for oil-base paints may be temporarily left in turpentine substitute or white spirit in jars, but they should always be supported vertically with the hairs clear of the bottom and the level of the spirit should be above the top of the hairs but not above the metal ferrule. A simple wire clip can be made up to hold the brush in this manner, as in Fig. 98.

But if a brush that has been used on oil paints is not required for further use for some time, it is much better to thoroughly wash it in luke-warm water and soap *after* a thorough wash in white spirit, after which it can be dealt with as described previously.

Brushes which have been used on cellulose paints should be thoroughly washed in alcohol, methylated spirit, or cellulose thinners as soon as possible after painting has been completed. It is essential to prevent the quick-drying cellulose paint from drying on the hairs; in fact, the hairs may be occasionally dipped into methylated spirit while painting proceeds, to prevent this.

Two final points on brush painting. Do not use the paint directly from its original tin, but decant enough for the job in hand into a suitable receptacle just before starting. A small glass jar which is quite clean and dry is best. Then seal the balance of the paint in the tin as quickly as possible, before dust can get into it.

Do not try to use paints which have been in stock too long and have gone "bitty", or have formed a thick "skin" on top. No amount of stirring will cure this, and the only hope of using it is careful straining through fine muslin. But this is usually a very slow process and as the small tins of paint used on small gauge model locomotives are comparatively cheap, it is really much better to cut one's losses and purchase fresh paint.

Lettering

Many otherwise excellent models are spoilt by incorrect, mis-shapen or badly spaced lettering. Even if transfers are used, the spacing of the letters is often incorrect.

Dealing first with scale models of some particular full-size locomotive, the first thing to do is to find out the correct size, spacing and colouring of the letters and numerals required. The necessary information can often be obtained from books, such as those published by the Railway Correspondence and Travel Society. Help can also be obtained from the Historical Model Railway Society. Failing this, photographs, especially full side views, of the prototype should be obtained.

Having decided the size and spacing of the letters or numerals required, it is a good plan to cut out a piece of thin cardboard and set out to scale the letters in their correct relative positions. This can then be placed just above the position on the model where the actual letters are to be painted.

It is advisable to practise painting the letters on odd scraps of metal which have been painted in exactly the same manner as the model itself. In this way, the best consistency for the paint can be determined beforehand. An oil based paint is probably best for the purpose.

Much locomotive lettering is in gold with black and/or red shading. The builder is advised to first paint the letters in chrome-yellow, and after this is dry, apply a thin coat of gold paint over the yellow. This will prevent the gold paint from running outside the edges of the letters.

If transfers are to be used for the letters and numerals, these should be of the "varnish-fixing" type. Varnish transfers are always supplied on thin transparent paper and this is fixed to a somewhat thicker white paper, the function of which is merely to protect the transfer before use.

This thicker white paper is first removed and the varnish should then be applied to the letters only. The varnish should be a quick-drying type, known in the Trade as "quick-jobbing" varnish.

After a few minutes, the varnished surface of the letters or numerals will have become tacky and the transfer is then pressed down evenly on the surface of the model. The transfer is now left for about half-an-hour, after which the thin backing paper is washed off in cold water, but this should be done with great care. On no account should attempts be made to pull the paper off the surface, as this may pull the lettering, or a part of it, away from the model.

If no lining is required on the model, the lettering, when thoroughly dry, should be protected by a thin coat of varnish.

Lining

The lining of model locomotives is probably the most difficult of all the finishing operations. Once again, transfers may be available for lining, but the drawback of lining transfers is that only straight lines will be available.

A good quality draughtsman's spring-bow pen is the best instrument for lining models. This should be charged, not by dipping it into the paint, but by filling it with a proper pen filler, which can be made from a short length of fine brass tube, the end of which is cut off at an angle and polished smooth.

Some builders prefer water colours for lining work, others claim that oil-based paints give the best result. As with lettering, the builder should practise repeatedly on odd scraps of painted metal, not only to establish the best consistency of the paint, but also the best "setting" of the pen. Having adjusted the pen properly, care should be taken not to disturb the adjusting-nut, which should be a fairly close fit on its screw.

The tip of the pen must be quite smooth and clean before attempting to fill it with the paint, which should be thin enough to run through the pen with sufficient ease to give a clean line.

Before commencing lining, make sure that the model being lined is properly supported and unable to move during the operation.

Some builders put the curves or curved corners of the lining in first, the compass point being stuck into the painted surface or into a piece of celluloid held to the surface. A better way, which has been demonstrated to the author by Mr. H. A. Taylor of Bletchley, is to make up a complete set of templates for the lining, which is generally found on cab sides, tank sides, tank fronts, tender sides, tender backs, etc., etc.

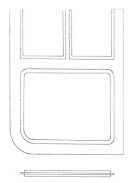

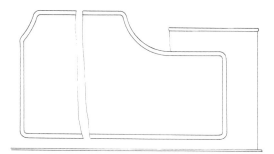

Figs. 99 and 99A Templates for lining work

These templates are cut out from flat steel sheet about $\frac{1}{16}$ in. thick; they must be exactly made and all the edges polished nice and smooth. On each side of the steel, pieces of cardboard, slightly smaller all round than the steel, are stuck by an adhesive such as Araldite or Loctite. Two lifting knobs are then fitted by tapping the steel, so that the knobs can be unscrewed and then screwed on the other side, when the other side of the model is being lined. It will be appreciated that the cardboard will hold the steel plate sufficiently away from the surface of the model to prevent "under-run" of the paint. The scheme is shown in Fig. 99.

The idea of using steel plate for these templates rather than brass or other non-ferrous metal is so that a small magnet (or two or more) may be placed at the back of the cab side or tender side, which will hold the template firmly in place while the lining is carried out. This of course cannot be done in the case of certain parts such as tank fronts, tender backs, etc., for which positions other means can be devised to hold the template.

If two or more lines are required, close together, the draughtsman's pen should be modified as shown in Fig. 100. A thin brass strip is attached to one side of the pen as shown, the thickness of the lower end being equal to the distance required between the lines. It may be necessary to fit a longer adjusting screw to the pen to allow for the thickness of this strip, and as the lower end of the pen will probably be hardened, the upper screw will have to be placed higher up where the metal will be softer.

Fig. 100 Modified draughtsman's pen

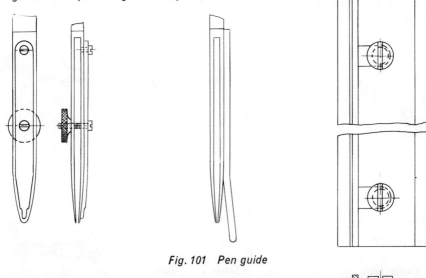

Fig. 101 Pen guide

Fig. 102 Jig for lining boiler bands

In use, the pen with the guide attached and adjusted is filled, and the inner line drawn all around the template. The pen is then reversed so that the guide bears against the template and the second line produced.

It might be thought that the presence of dummy rivet heads on the model would prevent the template from being held down properly; but depressions can be made in the cardboard to overcome this problem.

To line the running-board angle (valance), another type of guide should be made up and fitted to the pen. This is shown in Fig. 101. In this case, the guide is held against the bottom edge of the angle while the pen is moved along the model. If more than one line is required, the second line can be produced by fitting a guide with a greater gap between the guide and the centre of the tips of the pen.

After completing the lining on the running-board angle, the bottom edge of this will require touching up with the appropriate body colour, this being done with a fine brush.

Perhaps the most difficult parts of the model locomotive to line are the boiler bands. It is clearly a most tricky operation to line these in position on the model, owing to handrails and other details getting in the way of the pen. However, it is possible on models of the larger scales, and even gauge "1" models, to paint and line the boiler bands before fitting them to the boiler, using a small nut and bolt to secure the boiler band in position. This is of course similar to the method used on the full-size locomotive. For a gauge "1" model, the fixing screws would have to be not larger than 14 BA.

If the boiler bands are lined "in the flat", a jig could be made up, not only to hold the boiler band straight, but to act as a guide for the lining pen.

Name and Number Plates

On the full-size locomotive, the various name and number plates, manufacturers' plates, shed plates, etc., are generally brass castings, though sometimes iron castings were used.

The best way to produce these plates for small gauge models is to adopt the "photo-etching" process. The method is to first make a drawing of the plate to four times the actual size required, using Indian ink on clean white card (for gauge "1" plates, twice full size would be large enough). This drawing is then sent to a printers' block-maker, with the request to produce a "positive" on the required material, with the necessary reduction.

It will be appreciated that a normal printers' block would be unsuitable, as this would be a "mirror image" of the plate.

On receipt of the plates, they are cut out with a fine jewellers' or metal-cutting fretsaw, the edges cleaned with files, and the plates soldered in the required position. The "backgrounds" of the plates are then filled with paint of the desired colour, and finally the faces of the plates are carefully polished with fine emery.

These name and number plates will be found to give the final detail touches to an accurate scale model, and deserve the greatest care in the arrangement and finishing.

Index

Adams bogie   67
Aerosol paints   153
Air pumps   137–9
Air vents   109, 116, 117
Allan valve gear   76
American locomotives   142–7
Anchor link   79–80
Annealing   93
Armature brushes   38, 41–2
Armature spindle   34, 35, 38, 39, 43, 44, 148, 149
Armature stampings   30, 35, 36
Armature winding   39–41
Articulated locomotives   18–19
Ashpans   144
Axleboxes   65–6, 69, 70, 135
Axles   57–9, 68, 69
    coupled   21, 25, 27, 52, 65, 66, 142
    driving   21, 25, 27, 28, 52, 57–9, 65, 66, 142

Baker valve gear   61
Ball-bearings   35
Bars, compensating   69
Beading   124–6
Bells   146
Belpaire fireboxes   8, 96, 97
Blower pipes   118
Blower valve   126
Bogie frames   69–70
Bogies   12, 67–70, 115, 136
Boiler bands   99–100
Boiler mountings   100–4, 145
Boilers   8, 12, 19, 83, 91–5, 96–100, 105, 108, 143, 153
    taper barrels   8, 94–5, 118
Brake blocks   136–7
Brake pipes   121

Brakes   134–40
Brass   82, 83, 93
Brushes, paint   154–5
Brush gear   38, 41–2
Buffer beams   83, 84–85, 124
Buffers   82, 121–124
Bushes for driving and coupled axles   25, 27, 28, 29

Cab fittings   126–33
Cabs   83, 89–90, 105–6, 124, 125, 126–33, 154, 157, 158
    window beading   124–6
Chassis   20–6, 27, 32, 34, 142
Check valves   92, 118–20
Chimneys   82, 100, 101–4, 145–6, 154
Clack valves—see *Check valves*
Classes of locomotive   7–19
Coal   117, 150
Coal floor   112
Coal plate   110, 112
Coal rails   115–6
Commercial mechanisms   20–9, 30–44
Commutators   31, 36–9, 40, 41, 44
Compensating bars   69
Connecting rods   61–4
Correcting links   79–80
Coupled wheels   62
Coupling hooks   122, 124
Coupling rods   61–4, 136, 142
Couplings   114, 115, 124
"Cowcatchers"   143, 144–5, 146, 147
Crank axles   59–61
Crank bosses   56, 57
Crankpins   56, 57, 58, 59, 60, 61, 63, 64, 78, 142
Crossheads   73–6

Cross stretchers 22, 26, 27, 84
  bogie 69
Current collectors 26, 45–51, 65
  stud-contact system 49–51
  third-rail 45, 46, 47
  two-rail 45, 47, 48, 49, 65
Cylinder drain cock control 126
Cylinders, outside 70, 71–81

Domes 82, 83, 101–4, 116
  water pickup 116
Drag beams 83, 84–5
Drain cocks 73
Drilling machines 3, 4
Drills 4
Dummy bolt heads 71–2
Duralumin 83

Eccentric straps 79
Electrically driven models 20–9, 30–
  44, 45–51, 78, 81, 82, 126
Expansion links 76–78, 80
Express passenger trains 8–12

Firebox damper control 120
Fireboxes 8, 9, 12, 83, 91–5, 96–8, 143
Flat gears 148
Flexible drives 147–9
Fluting 62, 63, 64, 79
Flywheels 43–4
Freelance locomotive designs 19
Frame construction 20–9, 141–2, 143,
  150
  bogie 69–70
  tender 112–3
Frame plates 22, 26, 27

Garden railways 49
Gearboxes 28, 29, 67
Gearing 42–3
Gear ratios 43
Gears, valve—see Valve gears
Goods engines 14, 16, 43, 51
Guard irons 69

Handrail knobs 118
Handrails 83, 92, 118
Headlamps 146–7
Horn faces 135
Hornstays 66, 141

Insulating blocks 26

Jewellers' glass 6
Joy valve gear 76, 79–80

Lamp brackets 119
Lathes 3
Lettering 155–6
Lifting rings 109
Lining 156–9
Links
  anchor 79, 80
  correcting 79, 80
  vibrating 80
Locomotive classes 7–19

Machine tools 3–4
Magnets 21, 22, 29, 30–2, 34, 44, 67
Materials 82–3
Micrometers 4, 27
Mineral engines 14, 16, 43, 51
Mixed-traffic engines 13, 43, 82
Motor, making your own 30–44
Motors 12, 20, 21, 28, 30–44, 65, 66,
  143, 147, 148, 149, 150

Name plates 159
Nickel silver 81, 93
Number plates 159

Oil baths 43
Outside cylinders 9, 12, 70, 71–81
Outside-frame bogie 69–70
Outside-frame locomotives 141–2

Packing glands 72
Painting 151–6
Paints 151–3
Paint brushes 154–5
Pannier tanks 108–9, 139
Pantograph principle 50–1
Passenger engines, express 8–12
Pilots 143, 144–5, 146, 147
Pipes,
  brake 121
  steam heating 121
  water-feed 83, 118
Pole pieces 31, 32–35, 38
Pony trucks 70
Power tenders 147–50
Prototypes 7–19

"Quartering" 142

Radius rods 78–9, 80
Reamers 4, 54
Return cranks 61, 79
Reversing levers 126, 133
Riveting 139–40
Running boards 83, 142–3, 159
Running plates 83–4, 87–9, 142

Saddle tanks 106–8
Safety valves 82, 101, 118, 119
Screwing tackle 4
Side control springing 68
Side tanks 105, 106, 139
Single bar crossheads 75–6
Single wheeler locomotives 13
Slide bars 71, 73, 74–6
  securing 76
Slide gauges 4
Sliding blocks 68
"Slippers" 73–4, 75
Smokebox 83, 91, 92, 98–9, 139, 154
Smokebox doors 123
Soldering 1, 5, 6, 84–5, 100, 113
Soldering equipment 5, 6
Sound of wheels 65
Spectacle plates 87, 90, 98, 105
Spectacles 126

Splashers 83, 90, 124
Springing 65–70
  bogie wheels 68
  side control 68
  tenders 114, 150
Springs 124
Sprung driving axles 62, 65
Spur gears 149–50
Stainless steel 83
Steam brake valves 126
Steam heating pipes 121
Steam heat valves 126
Stephenson link gear 76, 79, 80
Steps 89, 120–1, 124
Stud contact system 49–51
Stuffing boxes 72
Superstructures 9, 12, 13, 22, 44, 82–101, 142

Tank engines 8, 16–18, 48, 105–9, 136, 139
Taper boiler barrels 8, 94–5, 118
Taps and dies 4
Tender frames 112–3
Tenders 48, 51, 110–7, 147–50, 157
  power 147–50
Tender springing 114–5
Tender tank filler 116
Tinplate 82
Tinsnips 84
Tool boxes, tender 112
Tools, construction 2–6

Union links 79

Vacuum brake pipes 121
Vacuum brake valve 126, 136
Vacuum gauges 126
Valances 83, 84–5, 86, 88, 159
Valve gears 61, 71–81
  Allan 76
  Baker 61
  Joy 76, 79–80
  Special types 80–1
  Stephenson 76, 79, 80
  Walschaerts 9, 13, 61, 72, 76–8, 79, 80